DATE DUE

D0760910

Calculus and Its Origins

The three aluminum pyramids together create a cube, demonstrating one of Buenaventura Cavalieri's arguments that appears in Chapter 4. (Photographs by Anthony Aquilina.)

Calculus and Its Origins

David Perkins
Luzerne County Community College

Published and Distributed by
The Mathematical Association of America

© *2012 by the Mathematical Association of America, Inc.*

Library of Congress Catalog Card Number 2011943235

Print edition ISBN: 978-0-88385-575-1

Electronic edition ISBN: 978-1-61444-508-1

Printed in the United States of America

Current Printing (last digit):
10 9 8 7 6 5 4 3 2 1

SPECTRUM SERIES

The Spectrum Series of the Mathematical Association of America was so named to reflect its purpose: to publish a broad range of books including biographies, accessible expositions of old or new mathematical ideas, reprints and revisions of excellent out-of-print books, popular works, and other monographs of high interest that will appeal to a broad range of readers, including students and teachers of mathematics, mathematical amateurs, and researchers.

MAA Service Center
P.O. Box 91112
Washington, DC 20090-1112
800-331-1622   FAX 301-206-9789

Preface

I was a born juggler, and after seven years of teaching college math students how to juggle, I was disenchanted. I had begun to feel like a painting teacher who taught color wheels but never let his students paint pictures, let alone engaged them in discussions of why art is meaningful or why artists differ in their styles. My students weren't complaining, really. Some of them were also born jugglers. The rest simply shouldered the burden, as if accustomed to the idea that math is no more than a kit full of tools.

So I scrambled my calculus class to place the discovery of calculus at the *end*. I intended to teach calculus as the culmination of an intellectual pursuit that lasted two thousand years. I could not find a suitable textbook, so I taught for three years using my own notes and homework problems. Then, on sabbatical, I wrote this book.

At first, I expected the book to be a synthesis of my notes and problems, but my research led me down new paths to many pleasant surprises. I had not realized the extent to which scholars in countries like Egypt, Persia, and India had absorbed and nourished Greek geometry when the western world went dark. Nor had I fully grasped how carefully ancient thinkers treated puzzles that lurk in the infinite. I gradually learned that because of these puzzles, calculus was not *discovered* in a way that would allow me to place its discovery at the "end" of anything.

Those we credit with the discovery explained the infinite in poetic terms. Even statements within the proofs themselves sound like metaphors: this tiny number is both zero and not-zero at the same time; this solid cube is composed of infinitely many flat slices. When pressed, mathematicians defended themselves with analogies. Isaac Newton used the example of a book to suggest how a three-dimensional volume could be composed of two-dimensional parts. Of course, a page is not strictly two-dimensional; it merely symbolizes such a thing. But we forgive him, not only because his calculations *work* but also because the rest of his arguments are insightful and rigorous.

Such flights of intellectual fancy do students a huge favor: the most mind-contorting technicalities are replaced by intuitive, appealing, simple arguments that are a pleasure to study. The figures alone speak eloquently about the subject. All that is required beyond algebra and basic geometry is a willingness to untether one's creativity when thinking about the infinite. And what student would be unwilling to do that in return for the chance to learn calculus as a pursuit rather than as a toolkit?

This subject is a treasure of the human intellect, pearls strung by mathematicians across both cultures and centuries. I hope this book holds a mirror up to this beauty.

Notes for professors

If you are a professor who assigns this book, I encourage you to ask your students to read a selection before class. I wrote with this model in mind. Intent and careful reading of professionally-written mathematics instructs students in their own writing. I ask my students to write in a journal as they read, so they can jot down questions, create their own examples, work out steps the author skipped, and re-create the figures as the author narrates. My students who take this seriously enjoy a noticeable upswing in the quality of their own writing. Further, I found that classes became far more dynamic, because the content of each class meeting was driven by the questions the students brought.

Each chapter concludes with a section entitled **Furthermore**. These sections introduce notable historical figures as well as a few results that are used in later chapters. The 'exercises' are designed to be read even if they are not assigned as homework. They are not meant to be difficult, but rather to be good checks on whether the reader has paid careful attention to the text.

I steered away from routine practice problems, such as lists of derivatives that require the product rule. I also elected not to weave much about the thinkers themselves into the narrative, for this information is widely available.

My thanks

Gerald Alexanderson gracefully served as managing editor from the beginning. Victor Katz's comments prompted a wholesale improvement of my discussion of the origins of the coordinate system in chapter 3. Christoph Nahr translated French and Padmini Rajagopal translated Malayalam for me. Ryan Walp gave me the aluminum cube (pictured in the frontispiece) as a gift, and Anthony Aquilina photographed it. Along with Ryan, my students Kelly, April, Shaun, Dave, and Eric read one of my drafts with me in an independent study class. Sam Fee gave me his photograph of the desert scene that appears on the front cover. I love this scene for the sand, which reminds us of the infinite, the desert, which is characteristic of the countries where calculus finds its origins, and the patterns created by the wind, which look not only like curves but also like notation. Michelle LaBarre did for this preface (and my life) what the wind did to the sand.

To Richard Jacobson,
who taught me mathematics, and told me
upon my graduation that if I wanted to
continue studying the subject, he would
support me. He thought we would get
along well, mathematics and I. I thought
about it for five years before I took him
up on his offer to kickstart my career.
This book would not exist without Jake.

Contents

1

The Ancients

A genie (as the story is told) lights a candle at a minute before midnight. After half of the minute has passed, the genie extinguishes the flame. Fifteen seconds later, she relights the candle, and again, halfway to midnight, she puts the flame out. This continues as midnight approaches, the time always divided in two, the flame soon leaping up and vanishing faster than we can see.

Now the genie asks you, "At midnight, will the flame be lit or out?"

Leaving aside the issue of *when* this question is asked, you are still left with some bewildering possibilities. The candle is neither lit nor out? The candle is both lit *and* out? We never *get* to midnight?

But of course we *get* to midnight; there has yet to be a midnight that we have failed to get to. There is a midnight right now that is approaching. Or are *we* approaching *it*? Which is staying still? Which is the arrow and which is the target?

One thing we have learned during the story of physics (in 1632) is that nothing sits still; you may see a passenger on a boat and a bird perched on the mast over her head as 'moving', but from their joint point of view, you are the one who is moving. Later in the story (in 1905), we learned that one observer may experience time as running more slowly than does another observer.

Questions like these, that explore the nature of reality, are ancient, and the pursuit of the answers has led thinkers down paths that we have named: physics (Can the genie's candle exist? Is time infinitely divisible?), philosophy (If our shared experience of time does not reflect its true nature, then what does that say about hope? knowledge? ethics?), and mathematics (Is it possible to make sense of all this?). This book approaches calculus as the culmination of a journey that began when people asked questions like these.

1.1 Zeno holds a mirror to the infinite

The question-asker **Zeno** lived (according to our linear experience of time) 2500 years 'ago' in Greece. One of his thought experiments questioned motion itself. We may phrase it like so: suppose an arrow, aimed at a target, is fired. Its tip must travel halfway to the target, then halfway again, and so on, reminiscent of the genie's candle, only with distance in place of time. This halving process continues indefinitely; thus, the arrow does not reach the target.

You are reading these words thanks to light bouncing off a surface and traveling to your eye. Why does the light not fall prey to Zeno's paradox? Can every distance, no matter how small, be divided in two?

Distance is tangible; we see it, we move through it. It feels infinitely divisible, just as time feels as though it is flowing into the future. It is tempting to discard reality and recast the paradox in terms of abstract objects that *are* always divisible: numbers. In this language, Zeno's tale becomes an infinite sum

$$\frac{1}{2} + \frac{1}{4} + \frac{1}{8} + \frac{1}{16} + \cdots$$

that equals 1 if the arrow reaches the target and never quite gets to 1 if the paradox holds.

There are several ways to argue that this sum never *exceeds* 1. But what if it never reaches 1? The sum, as we go along, is always growing; so, if the sum never reaches 1, it seems clear that it reaches something smaller than 1. This is a little like moving the target a bit closer to the archer and then asking Zeno what he thinks now. But we can show that no matter what target we choose smaller than 1, the infinite sum eventually slips past it into the gap between the target and 1.

For what if we stop the sum at some point and calculate the total thus far? After two terms, the total is $1/2 + 1/4 = 3/4$. (This number is called a *partial sum* of the infinite sum.) The next partial sum is $7/8$. In general, the partial sum after we add the first n terms is

$$1 - \frac{1}{2^n} . \tag{1.1}$$

The fraction $1/2^n$ represents how far from 1 the partial sum is when we stop adding at the nth term. So, no matter where we set the target, we can always choose n large enough (thus making $1/2^n$ small enough) to slip by.

When we say that the partial sums become "arbitrarily close" to 1, this is what we mean: there is no number smaller than 1 to which the partial sums approach. No, the partial sums approach 1. What conclusion can we draw other than

$$1 = \frac{1}{2} + \frac{1}{4} + \frac{1}{8} + \frac{1}{16} + \cdots \tag{1.2}$$

in the face of this argument?

If you believe that (1.2) uncomfortably stretches the notion of 'equals', then you keep good company. Nevertheless, it is true, especially in the story of calculus, that mathematics often advances thanks to people who are willing to take uncomfortable risks. Consider, for example, this approach to proving that (1.2) is true. If we assume that the infinite sum in (1.2) adds to some number that we call S, then we have

$$S = \frac{1}{2} + \frac{1}{4} + \frac{1}{8} + \frac{1}{16} + \cdots \tag{1.3}$$

and can multiply both sides by 1/2, yielding

$$\frac{1}{2}S = \frac{1}{4} + \frac{1}{8} + \frac{1}{16} + \cdots .$$

Subtracting the new equation from (1.3), we have

$$S - \frac{1}{2}S = \frac{1}{2} . \tag{1.4}$$

The vexatious infinite, as depicted by the three dots in (1.3), vanishes thanks to our ingenuity. Leaving aside the issue of how one might multiply an infinite number of terms by anything, we can conclude that $S = 1$. Figure 1.1 offers visual support for this conclusion. Might we leave aside this issue permanently, and treat infinite sums just as we do normal sums? That is an option, but if someone finds a troublesome example, we should pay attention. Infinite sums are often troublesome.

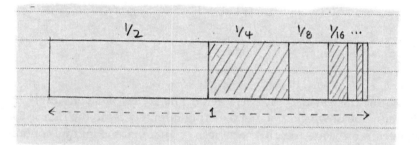

Figure 1.1. This figure speaks without words about the mathematics underlying Zeno's paradox of the arrow.

For instance, in the infinite sum

$$\frac{1}{2} + \frac{1}{3} + \frac{1}{4} + \frac{1}{5} + \cdots \tag{1.5}$$

each term is smaller than the last, so we might guess that this sum, too, adds to something. But it does not. Name any large number you want, and the partial sums will eventually pass it by. There is no cap on how large this sum grows, but over a thousand years passed after Zeno's life before someone could persuasively argue why.

By simply alternating the signs in (1.5), however, we get the infinite sum

$$\frac{1}{2} - \frac{1}{3} + \frac{1}{4} - \frac{1}{5} + \cdots$$

that approaches a finite number (approximately 0.693). Two thousand years after Zeno, someone identified this number. Puzzles that take centuries to solve warn us of deep mystery.

Because we usually think of adding when we say 'sum', we can use a more general term than 'infinite sum' that allows subtraction: *series*. The series

$$S = 1 - 1 + 1 - 1 + 1 - 1 + \cdots$$

sparked much debate. What might it equal? By grouping pairs of terms like so,

$$S = (1 - 1) + (1 - 1) + (1 - 1) + \cdots$$
$$= 0 + 0 + 0 + \cdots,$$

we conclude that S equals 0. Skipping the initial 1 and *then* grouping pairs like so,

$$S = 1 - (1 - 1) - (1 - 1) - \cdots$$
$$= 1 - 0 - 0 - \cdots,$$

we discover that the sum is 1. Grouping, applied in slightly different ways, leads to two different answers; now *this* is troublesome.

It gets better; the grouping

$$S = 1 - (1 - 1 + 1 - 1 + 1 - \cdots)$$

reveals the original series to *contain itself*, so

$$S = 1 - S,$$

leading to the answer $S = 1/2$. We again find a new answer simply by using parentheses (one of which is somehow flung infinitely far to the right). There are, in fact, even more ways to sum the series, a signal that the infinite contains mysteries that do not succumb to ordinary arithmetic.

None of this deterred mathematicians from tinkering; in fact, such mysteries probably provoked them. There is no harm in exploring, and these explorations put humankind on a path that led to the discovery of a new branch of mathematics.

1.2   The 'infinitely small'

Archimedes (Greece, born *c.* 287 BCE) investigated the infinite fearlessly. An inventor and astronomer, Archimedes seems to have taken his greatest joy in pure mathematics. Many Greek mathematicians found the study of geometry among the purest of pursuits; after all, where on Earth can we find a square or an equilateral triangle or a circle? These shapes are unattainable generalizations of what we see in the world, made completely theoretical by the very purity that makes them elegant. No one can draw a circle; one can only draw things that *look* like circles. But we can *imagine* circles.

A geometer, pondering a shape, immediately asks about its area. Imagine slicing a circle like a pizza (or the ancient Greek equivalent thereof) and standing the slices together with their tips up as in Figure 1.2. Because the circumference of the circle

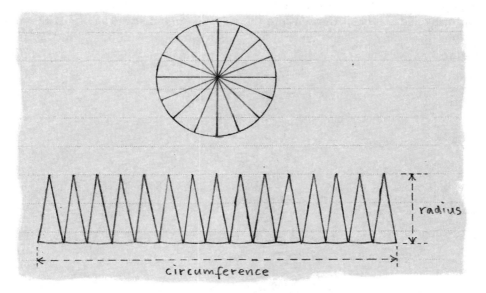

Figure 1.2. We rearrange the pieces of a circle to reveal the connection between its area, circumference, and radius.

has now been laid out in (virtually) a straight line along the bottom, the total area of these (pseudo) triangles sums to

$$\frac{1}{2}(\text{radius})(\text{circumference}) .$$

But this is the area of the original circle as well, so we suspect that the area A of a circle is related to the circumference C and radius r by

$$A = \frac{1}{2}rC . \tag{1.6}$$

This depends on how willing we are to believe that the thin slices are acting like triangles. Are you persuaded that the thinner the slices get, the more they behave like triangles? What if someone pointed out that as each slice gets thinner, its area becomes arbitrarily close to zero — in other words, the slice vanishes — so that if you want the slices to *be* triangles, they will first have to disappear?

This dispute notwithstanding, our argument leads to the correct conclusion: formula (1.6) tells the truth about circles. This blend of the infinitely many (the slices becoming more numerous) and the infinitely small (each slice on its way to vanishing) is at the heart of the story of calculus.

1.3 Archimedes exhausts a parabolic segment

Archimedes investigated the areas of circles (see exercise 1.1) and many other shapes, as well as the volumes of solids like cones and spheres. One of his efforts in particular carries us back to infinite sums. A *parabola* is, among other things, the path traveled by a ball thrown into the air, and Archimedes calculated the area within the *segment* created when a line cuts a parabola at points A and B, as in Figure 1.3.

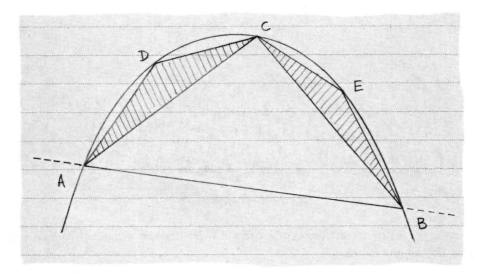

Figure 1.3. Archimedes exhausted the parabolic segment with successively smaller triangles.

Archimedes located[1] point C on the arc AB so that the line tangent to the parabola at C is parallel to AB; this creates triangle ABC. Letting $\triangle$ denote the area of this triangle, we proceed as he did, and create triangles ADC and CEB so that the tangents at D and E are parallel to AC and CB respectively. Archimedes showed that the areas inside these two new triangles totaled exactly one-fourth of the area of the first triangle ABC.

Now the area within the segment is pretty well exhausted by the three triangles; in fact, this method is called just that: *exhaustion*. How might he cope with the four small unfilled areas between the triangles and the parabola, however? Archimedes continued filling the unfilled areas with triangles, doubling the number each time, and proving that each new set of triangles totals one-fourth the area of the previous set. This process, continued indefinitely, produces the equation

$$\text{area within segment} = \triangle + \frac{1}{4}\triangle + \frac{1}{4}\left(\frac{1}{4}\triangle\right) + \frac{1}{4}\left(\frac{1}{4}\left(\frac{1}{4}\triangle\right)\right) + \cdots.$$

The equals sign is justified in the same way we discussed earlier: the partial sums grow arbitrarily close to the area within the segment as the host of triangles gradually exhaust that area. Factoring $\triangle$ from the right side yields

$$\text{area within segment} = \triangle + \triangle\left(\frac{1}{4} + \frac{1}{4^2} + \frac{1}{4^3} + \cdots\right),$$

---

[1]Although the arguments that underlie the geometry in this paragraph are missing, the reader should rest assured that such omissions are rare in this text.

which contains a sum much like (1.3). The first few partial sums are

$$\frac{1}{4} + \frac{1}{4^2} = \frac{4+1}{4^2} = \frac{5}{16} = 0.3125,$$

$$\frac{1}{4} + \frac{1}{4^2} + \frac{1}{4^3} = \frac{4^2+4+1}{4^3} = \frac{21}{64} \approx 0.3281,$$

$$\frac{1}{4} + \frac{1}{4^2} + \frac{1}{4^3} + \frac{1}{4^4} = \frac{4^3+4^2+4+1}{4^4} = \frac{85}{256} \approx 0.3320.$$

Further calculations suggest that the infinite sum equals $1/3$, prompting us to express the partial sums like so:

$$\frac{5}{16} = \frac{1}{3} - \frac{1}{3 \cdot 4^2},$$

$$\frac{21}{64} = \frac{1}{3} - \frac{1}{3 \cdot 4^3},$$

$$\frac{85}{256} = \frac{1}{3} - \frac{1}{3 \cdot 4^4}. \tag{1.7}$$

What we are deducting from $1/3$ in (1.7) is vanishing the further we go, so

$$\text{area within segment} = \triangle + \triangle \left(\frac{1}{3}\right) = \frac{4}{3}\triangle. \tag{1.8}$$

Because $\triangle$, the area of ABC, is simple to calculate, Archimedes had, in his own words, "shown that every segment bounded by a straight line and [a parabola] is four-thirds of the triangle which has the same base and equal height with the segment."

Despite his clever handling of ever-shrinking quantities, Archimedes carefully stated that he did not believe in numbers so small that they behaved like zero. (Specifically, he claimed that every positive number, no matter how tiny, may be added to itself enough times to create arbitrarily large sums.) As one of history's most accomplished mathematicians, Archimedes could tell when he was playing with fire, as we did when we generated (1.4). Centuries passed before anyone truly understood what he was being careful about.

1.4 Patterns

What do you see when you look at the drawing on the left in Figure 1.4? What if someone told you that this picture proved something lovely about odd integers? Pictures like this appear in documents that survive from several ancient cultures — Greece, India, China, Japan — despite the fact that such knowledge is unlikely to have been shared between these peoples. This leaves little doubt that mathematics belongs in the same discussion as music, poetry, and art when it comes to what pursuits are innate in humans. In fact, Figure 1.4 seems to speak as to how mathematics and art can befriend one another. As another example, the truth of the identity

$$\frac{1}{3} = \frac{1}{4} + \frac{1}{4^2} + \frac{1}{4^3} + \frac{1}{4^4} + \cdots$$

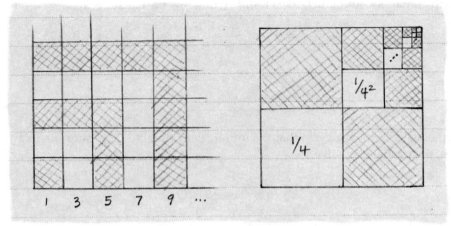

Figure 1.4. Each figure reveals a truth about a sum.

is revealed by the drawing on the right in Figure 1.4. The white boxes occupy areas equal to the terms on the right-hand side of the identity, but together they occupy one-third of the entire square. The drawing provides a more appealing argument, for some, than does our arithmetic approach in (1.7).

The image of a mathematical pattern often unlocks a secret. Consider the following pattern, known to all of the cultures mentioned above; starting at 1, add the positive integers consecutively, stopping at each partial sum:

$$1 + 2 = 3,$$
$$1 + 2 + 3 = 6,$$
$$1 + 2 + 3 + 4 = 10,$$
$$1 + 2 + 3 + 4 + 5 = 15,$$

and so on. Is there any pattern to the partial sums $3, 6, 10, 15, \ldots$ that we can use to predict something as difficult, say, as the sum of the first 1000 integers?

The partial sums are called *triangular numbers* thanks to the depiction in Figure 1.5. The nth triangular number T_n equals the sum of the first n natural numbers. We have the beginnings of an image that will help us find the 1000th triangular number without actually adding the first 1000 integers. Observe that any of the figures used to represent a triangular number can be copied and flipped, as in Figure 1.5. In each case, the nth triangular number has been copied to create an n by $n + 1$ rectangle. The number of squares in such a rectangle is simply $n(n + 1)$. Because T_n accounts for half the area of the rectangle, we are led to believe that

$$T_n = \frac{1}{2}n(n + 1) \, . \tag{1.9}$$

For example, T_4 is 10, and

$$1 + 2 + 3 + 4 = \frac{1}{2}(4)(5) = 10.$$

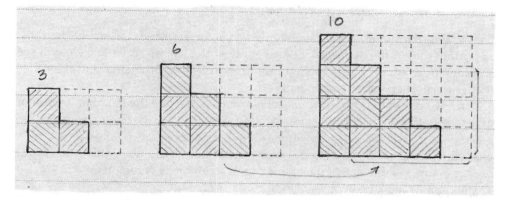

Figure 1.5. Each row of shaded squares represents an integer, and the triangular stacks represent their sums.

Using (1.9) is much simpler than adding the first thousand integers to find that $T_{1000} = 500500$.

This clever shortcut will appear obvious to many who have followed this argument, but, for the skeptics, we can argue the same point using language. The figures themselves suggest the path we will take. Look inside Figure 1.5 again. Hidden within the figure for $T_4 = 10$ is the figure for the previous case $T_3 = 6$, as indicated by the arrow. What if we knew that $1 + 2 + 3$ equaled half of 3 times the next integer 4; could we use this knowledge to show that $1 + 2 + 3 + 4$ equals half of 4 times the next integer 5?

Well, we *do* know that $1 + 2 + 3 = \frac{1}{2}(3)(4)$, because it is clearly true; just do the arithmetic. In fact, we can verify cases like this to our heart's content. But because there are rather too many cases to check in the long run, we will have to stop somewhere. Suppose we stop checking by hand after the first n integers, so that we know

$$1 + 2 + 3 + \cdots + n = \frac{1}{2}n(n + 1) .$$ (1.10)

If we can argue, in general, that adding the next integer $n+1$ preserves the pattern, will this convince you? Consider:

$$1 + 2 + 3 + \cdots + n + (n + 1) = \frac{1}{2}n(n + 1) + (n + 1)$$
$$= (n + 1)(\frac{1}{2}n + 1)$$
$$= \frac{1}{2}(n + 1)(n + 2) .$$

The first equality is justified by what we know from (1.10). The rest is factoring. Are you convinced that the sum of the first $n + 1$ integers is half of the integer $n + 1$ times the next integer $n + 2$? If so, then you are agreeing to believe in a method called *proof by mathematical induction*. This form of proof allows us to accept the evidence shown in the figures we have seen. Although ancient cultures used induction proofs implicitly, only in the 1600s would mathematicians formalize such arguments.

1.5   *The evolution of notation*

Underlying the patterns we have studied thus far are series, arguably the most important tool in the calculus kit. As we follow the story of calculus, we will see scholars grow adept at tackling ever more complicated problems, and the series they use will keep pace in complexity.

Thus, we pause to introduce *summation notation*, despite the fact that it was not invented until many centuries after Archimedes. (In fact, we have already transgressed in exactly this way by our use of signs like '+' and '−'. A Greek who wished to convey (1.10) would do so in words, and would not think of the numbers as quantities but as lengths of line segments or areas of regions.)

The capital Greek letter 'sigma' indicates a sum, as in

$$\sum_{k=1}^{n} k = \frac{1}{2}n(n+1) \, ,$$

which translates (1.10) into summation notation. The variable k is a placeholder that appears in the formula of the sum and increases from the number below Σ to the number above Σ by ones. We may write infinite series like (1.3) as

$$S = \sum_{k=1}^{\infty} \frac{1}{2^k} \, ,$$

and express our observations in (1.7) as

$$\sum_{k=1}^{n} \frac{1}{4^k} = \frac{1}{3} - \frac{1}{3 \cdot 4^n} \, .$$

The symbol ∞ denotes 'infinity,' indicating that the series never terminates. Again we rely on a symbol ∞ that entered general use centuries after Archimedes. We will use commonly known symbols for the sake of clarity, and summation notation thanks to its tie to series, but we will save the notation of calculus proper until the historical clock reads what it should.

1.6   *Furthermore*

1.1 **Archimedes estimates** π. Formula (1.6) is one of the tools needed to prove the well-known formula

$$A = \pi r^2 \tag{1.11}$$

for the area A of a circle in terms of its radius r and the constant π.

Quite a few ancient cultures knew that π was a number slightly larger than 3. Archimedes pursued this number by sandwiching it between two other numbers that he calculated using geometry. Taking a circle of radius 1 (and therefore of area π), he inscribed a regular hexagon within it (*regular* means that all of the angles are equal, and all of the sides are equal, as in Figure 1.6). Whatever the area of this hexagon, he could see that it equaled a number smaller than π.

(a) Find the area of the hexagon. (It is possible to do so using the Pythagorean Theorem, if you add a few lines to Figure 1.6.)

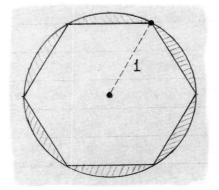

Figure 1.6. Archimedes began his approximation of π by inscribing a regular hexagon inside a unit circle.

(b) Archimedes also swapped the shapes, inscribing the circle of radius 1 in a larger regular hexagon. Find the area of this new hexagon. (This result is therefore larger than π.)

(c) To improve his estimates of π, Archimedes repeated both exercises using regular 12-sided polygons, which exhaust (and surround) the circle more completely. Find the area of such a 12-sided polygon that is inscribed in a circle of radius 1.

	# of red sides showing							
# of sticks tossed	0	1	2	3	4	5	6	7
3	1	3	3	1	-	-	-	-
4								
5								
6								
7								

Table 1. Each cell of the table contains the number of ways of tossing a certain number of sticks and getting a certain number of red sides.

1.2 **Binomial coefficients.** The boardgames Senet (Egypt) and The Royal Game of Ur (Mesopotamia) predate Zeno and Archimedes by well over a thousand years. In some versions of these games, the players tossed sticks rather than dice; each stick was two-sided, like a small popsicle stick, one side painted red, the other painted white. The number of red sides showing indicated the number of pieces the player could move.

For example, say that a player tosses three sticks. With the sticks labeled *A*, *B*, and *C* for convenience, we observe that there are three ways that the sticks will allow a player to move two pieces: the red sides can show on sticks AB, sticks AC, or sticks BC. Following this reasoning, we can fill in the first row of Table 1.

(a) Fill in the rest of the table.

(b) As you do so, look for any patterns you see, and express your observation in words.

(c) Argue on behalf of your observation in 1.2(b). One way to do this is to consider the fate of one of the sticks: is it white, or is it red? This breaks your counting problem into two cases.

> *Historical note.* The patterned numbers that appear in Table 1 were known, at least in part, to such scholars as **Omar Khayyám** (Persia, born 1048) and **Chu Shih-Chieh** (China, born *c.* 1260), to name just two. **Blaise Pascal**, who we will encounter again (exercise 6.1), revealed many truths hidden in these numbers, which are commonly named after him. These particular numbers have been of interest to scholars for many centuries; we return to them in exercise 3.4.

1.3 **The sum of the first n cubes.** A triangular number T_n is the sum of the first n integers and is calculated using formula (1.9). In this exercise, we explore the surprising connection between triangular numbers, cubed numbers, and squares.

(a) In particular, we seek a formula for the sum of the first n cubed numbers

$$\sum_{k=1}^{n} k^3 = 1^3 + 2^3 + 3^3 + \cdots + n^3 .$$

Calculate the sum for $n = 1, 2, 3, \ldots$ until you see a pattern in the sums that uses triangular numbers.

(b) Use a proof by induction to establish the truth of your pattern.

(c) In Figure 1.4, each image is designed to reveal a truth about numbers. Create an image that illustrates the connection that you have discovered about the sum of the first n cubes and the corresponding triangular number T_n. One such approach considers a cubed number as the product of that number times its square (or $m^3 = m \cdot m^2$ in symbols), and the result is a two-dimensional image.

1.4 **The sum of the first n squares.** Here we investigate another formula known to early cultures, including the Greeks. Write out the details of each step of this approach.

(a) We seek a formula for

$$\sum_{k=1}^{n} k^2 = 1^2 + 2^2 + 3^2 + \cdots + n^2 .$$

Calculate the sum for $n = 1, 2, 3, 4$. Because the formula we seek is a bit more complicated than most we have encountered in this chapter, it is not likely that these examples alone will suggest a pattern to you.

(b) Show that
$$k^3 - (k-1)^3 = 3k^2 - 3k + 1 . \tag{1.12}$$

(c) Substitute $1, 2, 3, \ldots, n$ for k in (1.12) and add all n of these equations together. (Why substitute for k, you may wonder? We are using k simply as a place holder; it is n that we care about, as we are seeking to add the first n squared numbers.)

(d) If we algebraically manipulate the result from part (c), we can arrive at the formula

$$\sum_{k=1}^{n} k^2 = \frac{1}{6}n(n+1)(2n+1)$$

$$= \frac{n^3}{3} + \frac{n^2}{2} + \frac{n}{6} .$$

Show this work.

2

East of Greece

My secret was not to listen
when my friend told me
that the stars answered all our questions.
He died beside the telescope
on a night he'd scribbled
"Saturn, Venus, Mars, aligned."
— *Ibn Khatir Tells How He Survived the Black Death,*
Thom Satterlee (2006)

The rise of Rome at the expense of Greece marked a steep decline in the pursuit of pure mathematics in the western world. Although Roman culture borrowed freely from Greek religion, philosophy, and art, Roman mathematics largely confined itself to what was necessary for commerce and engineering. However, the economies and militaries of both Greece and Rome extended east as far as India, prompting trade not only of goods but of knowledge. Islamic versions of universities attracted thinkers and collected knowledge for the sake of science, acting as transfer points where thinkers carried mathematical ideas between cultures.

The scholar **Abu Ali al-Hasan ibn al-Hasan ibn al-Haytham** (born 965) acted as such a conduit of knowledge. Born in what is now Iraq, ibn al-Haytham traveled to Egypt to work on a river-control project, interacting with scholars familiar with Greek mathematics. Ibn al-Haytham studied light, particularly its role in eyesight, and produced beautiful results concerning surfaces, reflection, angles, and numbers. In this chapter, we will study one of his arguments that resembles later European methods to the point where it is tempting to believe that his ideas filtered north from Egypt during the intervening 700 years.

2.1 Ibn al-Haytham sums the fourth powers

Ibn al-Haytham's result relies on an extension of exercises 1.3 and 1.4, where we found formulas for the sums of squares and cubes:

$$\sum_{k=1}^{n} k^2 = 1^2 + 2^2 + 3^2 + \cdots + n^2 = \frac{n^3}{3} + \frac{n^2}{2} + \frac{n}{6}, \tag{2.1}$$

$$\sum_{k=1}^{n} k^3 = 1^3 + 2^3 + 3^3 + \cdots + n^3 = \frac{n^4}{4} + \frac{n^3}{2} + \frac{n^2}{4}. \tag{2.2}$$

Ibn al-Haytham persisted, proving that

$$\sum_{k=1}^{n} k^4 = 1^4 + 2^4 + 3^4 + \cdots + n^4 = \frac{n^5}{5} + \frac{n^4}{2} + \frac{n^3}{3} - \frac{n}{30}. \tag{2.3}$$

We supplement his inductive, purely verbal argument with the pleasing geometric interpretation shown in Figure 2.1. The heights of the rectangles climbing up the left-hand side are all 1 unit, but the widths of those growing across the bottom are the cubes of successive integers (and thus, for convenience, the scale of the widths is only suggestive). The area of each rectangle appears within it; note the fourth powers appearing as areas.

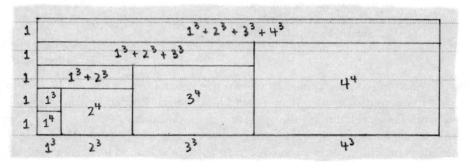

Figure 2.1. Ibn al-Haytham equated the entire area of the figure with the sum of its rectangular parts.

Consider the total area of the figure in two ways: as a simple height-times-width, and as the sum of all the areas of the rectangular pieces. Thus,

$$\begin{aligned}
(4+1)(1^3 + 2^3 + 3^3 + 4^3) = {}& 1^4 + 2^4 + 3^4 + 4^4 \\
& + 1^3 \\
& + 1^3 + 2^3 \\
& + 1^3 + 2^3 + 3^3 \\
& + 1^3 + 2^3 + 3^3 + 4^3.
\end{aligned} \tag{2.4}$$

Within this equality, on the right-hand side, appears the sum of fourth powers that ibn al-Haytham desired. On the left, we express the height of the large rectangle as $4 + 1$ rather than 5 so that later we can more easily substitute a variable like n for each 4 (where appropriate).

Using (2.2) to rewrite the right-hand side of (2.4) yields

$$
\begin{array}{ccc}
1^4 + 2^4 + 3^4 + 4^4 & & 1^4 + 2^4 + 3^4 + 4^4 \\[6pt]
+ 1^3 & & + \dfrac{1^4}{4} + \dfrac{1^3}{2} + \dfrac{1^2}{4} \\[10pt]
+ 1^3 + 2^3 & = & + \dfrac{2^4}{4} + \dfrac{2^3}{2} + \dfrac{2^2}{4} \\[10pt]
+ 1^3 + 2^3 + 3^3 & & + \dfrac{3^4}{4} + \dfrac{3^3}{2} + \dfrac{3^2}{4} \\[10pt]
+ 1^3 + 2^3 + 3^3 + 4^3 & & + \dfrac{4^4}{4} + \dfrac{4^3}{2} + \dfrac{4^2}{4}.
\end{array}
$$

On the right, summing the columns (excluding the top line) gives us

$$
1^4 + 2^4 + 3^4 + 4^4 + \frac{1}{4}(1^4 + 2^4 + 3^4 + 4^4)
$$
$$
+ \frac{1}{2}(1^3 + 2^3 + 3^3 + 4^3) + \frac{1}{4}(1^2 + 2^2 + 3^2 + 4^2),
$$

so we may replace the right-hand side of (2.4) with this result. Simplifying, we obtain

$$
\frac{5}{4}(1^4 + 2^4 + 3^4 + 4^4)
$$
$$
= \left(4 + \frac{1}{2}\right)(1^3 + 2^3 + 3^3 + 4^3) - \frac{1}{4}(1^2 + 2^2 + 3^2 + 4^2).
$$

Imagine that the rectangle in Figure 2.1 has $n+1$ rather than $4+1$ layers; we would have no trouble extending the figure in this way. Just so, we may replace 4 with n where appropriate:

$$
\frac{5}{4}(1^4 + 2^4 + \cdots + n^4)
$$
$$
= \left(n + \frac{1}{2}\right)(1^3 + 2^3 + \cdots + n^3) - \frac{1}{4}(1^2 + 2^2 + \cdots + n^2).
$$

One final use of (2.1) and (2.2) results in

$$
\sum_{k=1}^{n} k^4 = \frac{4}{5}\left[\left(n + \frac{1}{2}\right)\sum_{k=1}^{n} k^3 - \frac{1}{4}\sum_{k=1}^{n} k^2\right]
$$
$$
= \frac{4}{5}\left(n + \frac{1}{2}\right)\left(\frac{n^4}{4} + \frac{n^3}{2} + \frac{n^2}{4}\right) - \frac{1}{5}\left(\frac{n^3}{3} + \frac{n^2}{2} + \frac{n}{6}\right),
$$

which simplifies to (2.3) as promised.

2.2   Ibn al-Haytham's parabolic volume

The sum of fourth powers played a pivotal role in ibn al-Haytham's calculation of the volume created by revolving a two-dimensional shape around one of its sides.

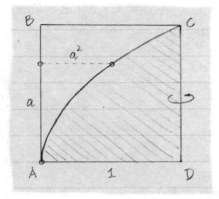

Figure 2.2. The shaded region revolves around CD to create the volume that ibn al-Haytham investigated.

The shape he revolved is shaded in Figure 2.2. Within square $ABCD$ (with sides 1) we draw curve AC so that for each distance a that we move vertically from A we then move distance a^2 horizontally.[1] Region ACD is revolved around side CD to create a volume much like a solid upside-down bowl.

Ibn al-Haytham sliced the area within region ACD into thin rectangular strips and asked what volume each strip would create when revolved around CD to create a thin cylinder. Totaling the volumes of these stacked cylinders, he estimated the volume within the upside-down bowl. Thinner slices give better estimates.

Let's say that there are n slices as in Figure 2.3, so that the height h of each slice is $1/n$. In the figure, only the kth slice from the bottom is drawn. We intend to revolve this slice around CD to create a cylinder. The volume of the cylinder requires the width. Remember how each point on curve AC is created: we move up from A toward B, then move the square of that distance to the right. So the gap between AB and the left-hand edge of slice k equals $(kh)^2$. The width of slice k is therefore $1 - (kh)^2$. Thus, the cylinder created by revolving slice k around CD has volume

$$h\pi\left(1 - (kh)^2\right)^2.$$

Because this cylinder does not perfectly fill its layer in the volume we seek, it only approximates the volume of its layer. No matter; later we will use thinner cylinders (by increasing n) to shrink the error.

On the left is the total volume of the n cylinders, and on the right we have expanded each square:

$$
\begin{array}{ccc}
h\pi\left(1 - (1h)^2\right)^2 & & h\pi\left(1 - 2(1h)^2 + (1h)^4\right) \\
+\, h\pi\left(1 - (2h)^2\right)^2 & & +\, h\pi\left(1 - 2(2h)^2 + (2h)^4\right) \\
+\, h\pi\left(1 - (3h)^2\right)^2 & = & +\, h\pi\left(1 - 2(3h)^2 + (3h)^4\right) \\
\vdots & & \vdots \\
+\, h\pi\left(1 - (nh)^2\right)^2 & & +\, h\pi\left(1 - 2(nh)^2 + (nh)^4\right).
\end{array}
$$

[1]In modern notation, the equation of curve AC is $x = y^2$ if A is $(0,0)$ and C is $(1,1)$.

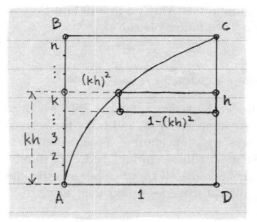

Figure 2.3. The kth slice has length $1 - (kh)^2$ and is revolved around CD to create a thin cylinder.

Collecting like terms down the right-hand side gives

$$h\pi\left[n - 2h^2\left(1^2 + 2^2 + \cdots + n^2\right) + h^4\left(1^4 + 2^4 + \cdots + n^4\right)\right]$$

$$= \pi\left[1 - \frac{2}{n^3}\sum_{k=1}^{n}k^2 + \frac{1}{n^5}\sum_{k=1}^{n}k^4\right].$$

Now (2.1) and (2.3) deliver the desired result:

$$\pi\left[1 - \frac{2}{n^3}\left(\frac{n^3}{3} + \frac{n^2}{2} + \frac{n}{6}\right) + \frac{1}{n^5}\left(\frac{n^5}{5} + \frac{n^4}{2} + \frac{n^3}{3} - \frac{n}{30}\right)\right]$$

$$= \pi\left[1 - \left(\frac{2}{3} + \frac{1}{n} + \frac{1}{3n^2}\right) + \left(\frac{1}{5} + \frac{1}{2n} + \frac{1}{3n^2} - \frac{1}{30n^4}\right)\right]. \tag{2.5}$$

Conveniently, the variable n appears only in the denominators of these fractions. It is safe now to let the slices become arbitrarily thin as n becomes arbitrarily large, and simultaneously each fraction with n in its denominator goes to zero. We conclude that the volume of the upside-down bowl is $(8/15)\pi$.

Ibn al-Haytham did not reach this conclusion quite as we did; he sandwiched the volume between two amounts that converged on $(8/15)\pi$ like two hands coming together to clap. These two amounts related to the volume of the cylinder created by revolving the entire region $ABCD$ in Figure 2.3 around CD. In our case, that volume is exactly π, and ibn al-Haytham actually proved that any parabolic bowl formed by the revolved curve $x = cy^2$, where c is a positive constant, will occupy 8/15 of the volume of the cylinder that encloses it.

This approach to finding a volume by cutting it into ever-thinner slices predates the official discovery of calculus in the 1600s by centuries. In fact, were ibn al-Haytham the direct ancestor of those discoverers, he would be their great-grandfather twenty times over. Currently, we have little evidence that ibn al-

Haytham's methods found their way north into Europe; but the transmission of ideas like his to the *east* is far more plausible.

2.3 *Jyesthadeva expands* $1/(1+x)$

India marked the most eastern conquest of the Greeks under Alexander the Great (the century before Archimedes lived) and trade between India and the Islamic nations to the west was brisk for centuries. For these and other (mainly political) reasons, the mathematics of these cultures had opportunity to blend. Early on, Indian mathematicians became aware of Greek investigations in geometry and astronomy. In turn, their discoveries found their way west as scholars traveled and translated.

Greece gave India the gift of geometry, and India gave the same gift back multiplied in value. A book written by **Jyesthadeva** (born *c.* 1500), for example, contains a sum that introduces yet another wonderful interaction between geometry and the infinite. Like ibn al-Haytham's before him, Jyesthadeva's argument pre-dates those of Europeans by over 100 years.[2]

Jyesthadeva's proof relied on a preliminary result that is itself an elegant blend of geometry and the infinite. He began with the identity

$$\frac{1}{1+x} = 1 - x\left(\frac{1}{1+x}\right),$$

which can be checked by simplifying the right-hand side. Jyesthadeva pointed out that the expression inside the parentheses matches the expression on the left-hand side, so he replaced the former with the right-hand side repeatedly:

$$
\begin{aligned}
\frac{1}{1+x} &= 1 - x\left(\frac{1}{1+x}\right) \\
&= 1 - x\left(1 - x\left(\frac{1}{1+x}\right)\right) \\
&= 1 - x + x^2\left(\frac{1}{1+x}\right) \\
&= 1 - x + x^2\left(1 - x\left(\frac{1}{1+x}\right)\right) \\
&= 1 - x + x^2 - x^3\left(\frac{1}{1+x}\right) \\
&= \cdots \\
&= 1 - x + x^2 - x^3 + x^4 - x^5 + \cdots .
\end{aligned}
\tag{2.6}
$$

Jyesthadeva thereby turned a simple fraction into an infinite series. We should be cautious; a cavalier treatment of the infinite can lead to absurd results. For exam-

---

[2]It is possible that some or all of his argument was discovered previously by fellow Indian **Kerala Gargya Nilakantha** (born *c.* 1450), but here we simply attribute all arguments to Jyesthadeva.

ple, substituting $x = 1$ in (2.6) leads to the peculiar

$$\frac{1}{2} = 1 - 1 + 1 - 1 + 1 - 1 + \cdots ,$$

which we encountered in Chapter 1. Or substituting $x = -1$ in (2.6) suggests the fantastical yet reasonable-looking

$$\frac{1}{0} = \infty .$$

Letting $x = 0$ in (2.6) results in the entirely boring but reassuring $1 = 1$. Choosing $x = 1/2$ gives

$$\frac{2}{3} = 1 - \frac{1}{2} + \frac{1}{4} - \frac{1}{8} + \frac{1}{16} - \frac{1}{32} + \cdots ,$$

which may or may not be true. So for what values of x does Jyesthadeva's result (2.6) lead us to truth, or to mystery, or to nonsense?

A bit more experimentation with various values of x in (2.6) will convince you that if x is between 0 and 1, we have hope. For now, we assume that x is such a number, and (although Jyesthadeva did not) turn to geometry for further insight. In Figure 2.4, square $ABCD$ has sides of length 1. Locate point E on BC so that $BE = x$. Let X be the point where segments BD and AE intersect.

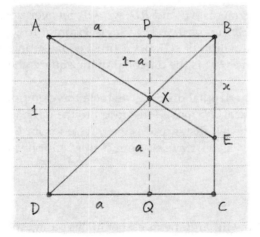

Figure 2.4. The importance of point X relies on a equaling $\dfrac{1}{1+x}$.

To identify the location of this special point, draw PQ through X parallel to AD and let $a = DQ = QX = AP$. The similarity of triangles APX and ABE gives

$$\frac{AP}{PX} = \frac{AB}{BE} \implies \frac{a}{1-a} = \frac{1}{x} \implies a = \frac{1}{1+x} .$$

So point X is the central focus of this figure, for its location generates the left-hand side of (2.6), the equation we are studying.

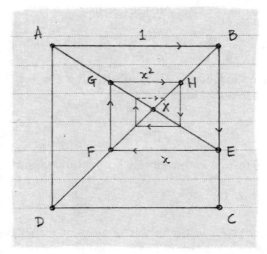

Figure 2.5. As we spiral in toward X from the starting point A, focus only on the horizontal displacement.

Using Figure 2.5 as a guide, imagine point X as the goal of a trip where we start at A, move to B, and then on to E, making another right-angled turn to reach point F on BD. We are getting closer to X; if we keep making right-angled turns each time we encounter segments AE and BD, we will get ever closer and closer.

Our next turn, at F, brings us to G. How far is it from F to G? Triangles ABE and EFG are similar, so the ratio of GF to FE must equal x, and FG must therefore have length x^2. Because BD is the diagonal of the square, the distance from G to H is also x^2.

Focus only on the horizontal distance we have traveled from the starting point A, and ignore the vertical. We moved distance 1 to the right as we traveled from A and B, then back to the left distance x as we traveled from E to F, and then back to the right distance x^2 as we traveled from G to H. Our horizontal displacement from A is

$$1 - x + x^2$$

when we are at H. Now the arguments earlier in this paragraph continue to supply us with terms as we spiral ever closer to X, justifying our belief that

$$\frac{1}{1 + x} = 1 - x + x^2 - x^3 + x^4 - x^5 + \cdots \tag{2.7}$$

when x is between 0 and 1. Thus is Jyesthadeva's algebra in (2.6) supported by geometry.

A series like (2.7), where the ratio of each term to the previous term is fixed (in this case, it is $-x$), is called a *geometric series*. On page 132 we will see a modern treatment of such a series.

2.4　Jyesthadeva expresses π as a series

Jyesthadeva's inventive expansion of $1/(1 + x)$ is the cornerstone of a more marvelous series, generated by the study of a circle's circumference. The quarter-circle in Figure 2.6 has radius 1, so its bisected arc AX has length equal to one-eighth of 2π, or $\pi/4$. By approximating arc AX with tiny line segments, Jyesthadeva discovered a series that equals $\pi/4$.

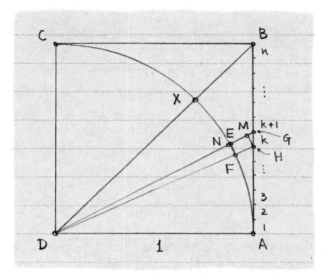

Figure 2.6. Jyesthadeva approximated the arc AX with many tiny straight pieces like FN.

He divided AB into n equal heights, each of height h. Focus on the kth such division, counting up from A. Drawing straight lines from D to each end of that division, we locate points E, F on arc AX and G, H on AB as shown. Pause to see where Jyesthadeva was heading: as we choose each division along AB from the bottom to the top, the arc AX is divided into n pieces. Although not of equal length, these pieces compose the arc AX that Jyesthadeva wished to measure.

The small arcs EF were not useful to Jyesthadeva because they are *exactly* what compose arc AX, and he already knew the length of that arc is $\pi/4$. So Jyesthadeva introduced two tiny segments, HM and FN, each perpendicular to DG. Figure 2.7 shows these details. Segment HM serves merely as a tool, but segment FN is almost indistinguishable from arc EF, especially as n grows larger. Getting our hands on FN, then, in terms of k and h, is the goal.

Jyesthadeva noted two pairs of similar triangles: triangles DFN and DHM (because FN and HM are perpendicular to DG) and triangles GHM and GDA (which share an angle at G and which each contain a right angle). Remembering that the quarter-circle centered at D has radius 1, we discover the relationships

$$\frac{FN}{DF} = \frac{HM}{DH} \implies FN = \frac{HM}{DH} \tag{2.8}$$

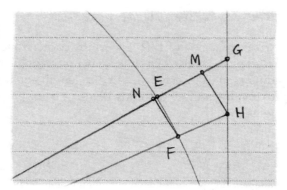

Figure 2.7. Segment FN approximates arc EF.

and

$$\frac{HM}{HG} = \frac{AD}{DG} \implies HM = \frac{HG}{DG} \, . \tag{2.9}$$

Substituting (2.9) into (2.8) gives

$$FN = \frac{HG}{DG \cdot DH} \, .$$

Lengths DG and DH differ by very little, and this difference approaches zero as we divide AB into more equal heights. So Jyesthadeva approximated FN with

$$FN \approx \frac{HG}{(DG)^2} = \frac{HG}{1 + (AG)^2},$$

using the Pythagorean Theorem for the last step. Finally, because AG is composed of k copies of $HG = h$, we have

$$FN \approx \frac{h}{1 + (kh)^2} \, . \tag{2.10}$$

This approximates EF as desired.

The length of arc AX in Figure 2.6 is $\pi/4$, and Jyesthadeva has now approximated that length as the sum of n tiny segments FN; further, his estimate

$$\frac{\pi}{4} \approx \sum_{k=1}^{n} \frac{h}{1 + (kh)^2} \tag{2.11}$$

improves as n grows larger. Now his expansion (2.7) plays its part. Because kh is the length of AG in Figure 2.6, and $AB = 1$, then $0 < kh < 1$ and so $0 < (kh)^2 < 1$. We need not worry about proceeding with the help of (2.7):

$$\frac{\pi}{4} \approx \sum_{k=1}^{n} h \, \frac{1}{1 + (kh)^2}$$

$$= \sum_{k=1}^{n} h \left(1 - (kh)^2 + (kh)^4 - (kh)^6 + (kh)^8 - \cdots \right)$$

$$= h \Big[1 - 1^2 h^2 + 1^4 h^4 - 1^6 h^6 + 1^8 h^8 - \cdots$$

$$+ 1 - 2^2 h^2 + 2^4 h^4 - 2^6 h^6 + 2^8 h^8 - \cdots$$

$$+ 1 - 3^2 h^2 + 3^4 h^4 - 3^6 h^6 + 3^8 h^8 - \cdots$$

$$\cdots$$

$$+ 1 - n^2 h^2 + n^4 h^4 - n^6 h^6 + n^8 h^8 - \cdots \Big].$$

Adding down the columns yields

$$h \Big[n - h^2 \sum_{k=1}^{n} k^2 + h^4 \sum_{k=1}^{n} k^4 - h^6 \sum_{k=1}^{n} k^6 + h^8 \sum_{k=1}^{n} k^8 - \cdots \Big]$$

$$= hn - h^3 \sum_{k=1}^{n} k^2 + h^5 \sum_{k=1}^{n} k^4 - h^7 \sum_{k=1}^{n} k^6 + h^9 \sum_{k=1}^{n} k^8 - \cdots$$

$$= 1 - \frac{1}{n^3} \sum_{k=1}^{n} k^2 + \frac{1}{n^5} \sum_{k=1}^{n} k^4 - \frac{1}{n^7} \sum_{k=1}^{n} k^6 + \frac{1}{n^9} \sum_{k=1}^{n} k^8 - \cdots . \qquad (2.12)$$

Aware of results (2.1) and (2.3), and how they extend to sixth powers and eighth powers and so on (see exercise 2.4), Jyesthadeva replaced the sums in (2.12):

$$1 - \frac{1}{n^3} \left(\frac{n^3}{3} + \frac{n^2}{2} + \frac{n}{6} \right)$$

$$+ \frac{1}{n^5} \left(\frac{n^5}{5} + \frac{n^4}{2} + \frac{n^3}{3} - \frac{n}{30} \right)$$

$$- \frac{1}{n^7} \left(\frac{n^7}{7} + \frac{n^6}{2} + \frac{n^5}{2} + \frac{n^3}{6} - \frac{n}{42} \right) + \cdots$$

$$- \frac{1}{n^9} \left(\frac{n^9}{9} + \frac{n^8}{2} + \frac{2n^8}{3} - \frac{7n^5}{15} + \frac{2n^3}{9} - \frac{n}{30} \right) + \cdots .$$

He simplified much as we did in (2.5), then let each fraction with an n in its denominator go to zero. What emerged is one of the world's most elegant formulas:

$$\frac{\pi}{4} = 1 - \frac{1}{3} + \frac{1}{5} - \frac{1}{7} + \frac{1}{9} - \cdots . \qquad (2.13)$$

Who would suspect this link between the constant π and a series involving the odd numbers?

It is one thing to read and understand the argument that produces (2.13) and another to conceive it. This impractical but lovely series was rediscovered by others who followed their own paths, and none of these people claimed to have pursued this goal for any reason other than the attainment of beauty. The charm of mathematics runs parallel to art, poetry, music, and literature; we can marvel at an analogy, a sculpture, a symphony, or a tale, and then mimic them if we desire. As is often the case with mathematics, the results are judged by their beauty.

2.5 Furthermore

2.1 **Brahmagupta seeks patterns in arithmetic.** One of the most ancient mathematics texts was authored by **Brahmagupta** (India, born 598), an astronomer whose interest in numbers stretched beyond what was merely practical. His serious treatment of zero illustrates this point; his bold claim that zero divided by zero equals zero foreshadowed a debate that we shall return to in chapter 9.

Brahmagupta devoted much attention to *arithmetic sequences*, which will play an important role in chapter 5. A sequence of numbers is *arithmetic* if the *common difference* between consecutive terms is constant. For example, the arithmetic sequence $5, 8, 11, 14, 17, 20$ has common difference 3. Its *initial term* is 5 and its *period* (that is, the number of terms in the sequence) is 6.

(a) Find the 100th term of the arithmetic sequence that begins

$$5, 8, 11, 14, \ldots .$$

(b) Find the nth term in the arithmetic sequence that begins

$$a, a + d, a + 2d, a + 3d, \ldots .$$

You may find it helpful to create your own numerical examples like the one in part (a), only simpler. These examples will not only nudge you in the correct direction as you seek the desired formula, but will also provide test cases for your conjectures.

(c) Brahmagupta claimed that anyone who knows the initial term a, common difference d, and period p of an arithmetic sequence could determine its sum S. What formula for S does the trick?

(d) Brahmagupta changed the 'givens' to pique the minds of his readers. Try this one: "On an expedition to seize his enemy's elephants, a king marched two *yójanas* the first day. Say, intelligent calculator, with what increasing rate of daily march did he proceed, since he reached his foe's city, a distance of eighty *yójanas*, in a week?"

(e) What if we know the sum S and period p of an arithmetic sequence, but lack the initial term a and common difference d; can we always find values for a and d that satisfy the givens? What is your answer?

2.2 **Porphyry, Pappus, and Bhaskara count.** The Syrian scholar **Porphyry** (born
c. 234), in his commentary on the works of **Aristotle** (Greece, born 384 BCE),
did not resist a detour into mathematics when the opportunity arose. Porphyry
wished to explain the differences between five 'qualities' that categorize all real
things (such as horses, the number 10, Aristotle himself, and so on). This am-
bitious undertaking prompted Porphyry to contrast each of the five qualities
with each of the others.

(a) Rather than simply begin, Porphyry paused to argue that there are ten such
    comparisons. Choose any of the five qualities; we may compare it to each
    of the other four. Now when we consider pairing a second quality to the
    others, we do not want to reconsider the first quality, which has already
    been compared to the second; rather, we pair the second to the remaining
    three. In this manner, Porphyry claimed, we see that the total number of
    pairings is $4 + 3 + 2 + 1 = 10$. (This is the fourth triangular number.)

    Use induction to argue that the number of ways to pair n objects is the
    triangular number T_{n-1}.

(b) The Greek mathematician **Pappus** (born *c.* 290) set his mind to many prob-
    lems of geometry, one of which led him to consider the same problem as
    Porphyry. Pappus counted the crossing points of lines that are drawn so
    that no two are parallel and no three intersect at the same point. Explain
    the connection between this problem and that of Porphyry.

(c) Centuries later, **Bhaskara** (India, born 1114) included results on counting
    in his summary of current mathematical knowledge. The problem of Por-
    phyry and Pappus generalizes naturally beyond pairs, and Bhaskara
    pursued this with his own colorful illustrations. His peers who mixed
    medicines, for example, classified the ingredients into six 'tastes' (sweet,
    sour, salty, bitter, astringent, and pungent). More than two tastes could be
    combined into one mixture.

    Bhaskara provided an easily imitated pattern to use in solving counting
    problems of this sort. To count the ways that we may mix n of the six tastes,
    multiply together n of the following fractions, starting on the left:

    $$\frac{6}{1}, \frac{5}{2}, \frac{4}{3}, \frac{3}{4}, \frac{2}{5}, \frac{1}{6}.$$

    For example, there are $(6/1) \times (5/2) \times (4/3) = 20$ ways to choose 3 of the 6
    tastes for the mixture. Why does Bhaskara's method work? You will have
    to explain how the denominators eliminate repetition.

2.3 **The series for** $1/(1-x)$. Taking Figure 2.8 as a starting point, add what lines and labels that are required to explain why

$$\frac{1}{1-x} = 1 + x + x^2 + x^3 + x^4 + \cdots$$

in the same way that we justified (2.7). You might start by finding the length of CF.

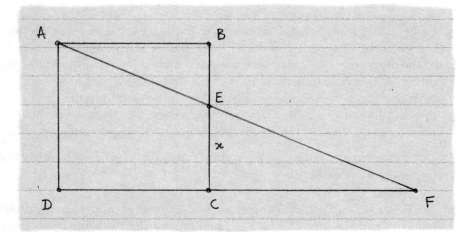

Figure 2.8. $ABCD$ is a square with sides 1, and both AEF and DCF are straight lines.

2.4 **Extending ibn al-Haytham's sums of powers.** Formulas (2.1) and (2.3) let ibn al-Haytham replace the first two sums in (2.12). To attack the ensuing sums, we can repeat the process suggested by Figure 2.1 for a few cases, and then draw a universal conclusion from our results.

(a) Follow ibn al-Haytham's path in section 2.1 to find the formula that sums the fifth powers:

$$\sum_{k=1}^{n} k^5 = \frac{n^6}{6} + \frac{n^5}{2} + \frac{5n^4}{12} - \frac{n^2}{12}.$$

(b) Inspect your work for a reason to believe that for any positive integer $p > 5$,

$$\sum_{k=1}^{n} k^p = \frac{n^{p+1}}{p+1} + \text{(terms of lower degree)}.$$

3

Curves

Say you bounce a ball
Have you ever noticed that
Between the business of its going up
and the business of its fall
it hesitates?

It just waits
There's a fraction of a second there
when it's luxuriating in the air
Before its fate rushes it on.
     — from "Circe and the Hanged Man", Ellen McLaughlin (2010)

Even if you sequester yourself in nature, away from the influences of humankind, the world moves in curious, patterned ways. Why does one falling leaf drift in a spiral while another twirls about its axis? What explains the eddy patterns in a creek? What forces act on a bird's wing? Why do the stars travel a circle during the night? Thinkers from many cultures looked beneath the surface of questions like these; underlying all of the answers was mathematics.

3.1 *Oresme invents a precursor to a coordinate system*

Most thinkers who discover something about how the universe works want to share their discoveries, and some created notation intended to streamline this communication. Chapter 7, for example, is in part devoted to explaining the symbols that we use today in calculus. Equally important in our story is the development of visual tools that, paired with sophisticated notation, not only facilitated the sharing of results but also sparked ideas that almost certainly would have otherwise remained out of reach. One such groundbreaking tool in mathematics is the coordinate system — a graphical way of picturing how two or more variables relate — with origins that reach back to medieval Europe.

If you rest one end of a metal poker (a long rod used to manipulate hot objects) in a fire and wait a bit, the temperature of the poker changes so that it is great at one end and less at the other — and we all know which end is which. **Nicole Oresme** (France, born *c.* 1323) suggested that the best way to depict this is to draw a horizontal line segment to represent the poker and, perpendicular to each point on the line, a vertical segment that illustrates how hot the poker is at that point. Figure 3.1 shows what he means for two pokers, one that has an end in the fire and another that is not near a fire. Oresme forthrightly stated that there could be "no more fitting way" to express such a physical concept in the form of a diagram.

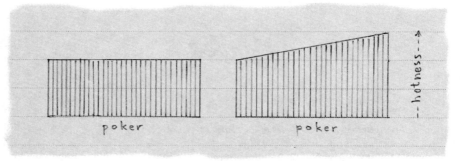

Figure 3.1. Do you agree that it is easy to determine which end of which poker is in a fire?

Today, we are used to assigning a scale to the vertical heights, but Oresme's focus was on the ratio of the heights; if, at one point, the height is double that of another, then the first point of the poker is twice as hot. It did not matter to Oresme what the two temperatures actually were. Further, Oresme elected to omit the vertical lines themselves and keep only the uppermost point of each, because, after all, there are infinitely many points along the poker. He let "the line of summit" refer to this collection of points. Figure 3.2 shows Figure 3.1 as Oresme drew it (along with a third poker).

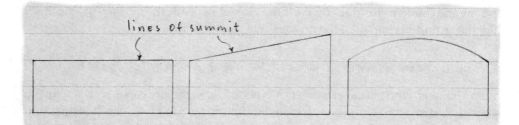

Figure 3.2. What does the third diagram imply about the location of the fire and the poker?

Oresme celebrated not only the clarity but also the versatility of these diagrams. If the object in question is two-dimensional, like a metal plate, then the vertical segments create a surface rather than a line. Further, the horizontal segment need not portray something physical; it can represent, for example, a period of time.

When we partner this last idea — that the horizontal segment can be a visual metaphor for the passing of time — with the choice to let the vertical lines represent the velocity of an object during that period of time, then we have come close to the heart of the host of questions that calculus answers. Pairing time with velocity, rather than points along a poker with hotness, we see in the first diagram of Figure 3.2 an object moving at a constant velocity, in the second an object undergoing an increase in velocity, and in the third an object that speeds up to a maximum velocity and then slows down to its original velocity.

Acceleration is the word associated with a change in velocity, made tangible by the force we feel when sitting in a plane that is taking off or on a horse that is skidding to a stop. In a plane or on a horse that is traveling at a constant velocity (the first diagram), we do not feel this force, indicating a lack of acceleration. By contrast, the second diagram represents an object with a constantly increasing velocity, and so what is called a *uniform* acceleration.

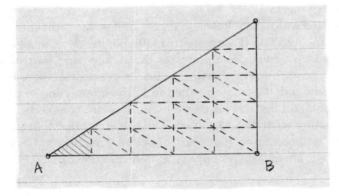

Figure 3.3. Subdividing the uniform acceleration diagram reveals a link to the odd integers.

Oresme took special note of the case of uniform acceleration starting at rest, which gives rise to the diagram in Figure 3.3. At time *A*, the object begins to move, and its velocity increases uniformly. Oresme observed that if we subdivide *AB* into *n* equal parts, then the small shaded triangle at *A* is repeated throughout the rest of the diagram, as shown. Altogether, the entire triangle consists of

$$1 + 3 + 5 + \cdots + (2n - 1)$$

such small triangles. Oresme knew (as we do from Figure 1.4) that this sum in n^2. Moreover, because distance is the product of velocity and time, Oresme connected this sum to the distance traveled by the object. So in the case of uniform acceleration, there appears to be a link between distance and the *square* of how much time has passed.

Three centuries later, **Galileo Galilei** (Italy, born 1564) stated this connection plainly:

> The spaces described by a body falling from rest with a uniform acceleration motion are to each other as the squares of the time intervals employed in traversing those distances.

Galileo experimentally verified this mathematical observation about the world, and today we express this fact with the formula $d = kt^2$.

When acceleration is not uniform, the analysis is more complex, and a full exploration only took place in the decades after Galileo's death in the mid-1600s. The rest of this chapter concerns two thinkers who not only engaged this problem of curved lines of summit but also refined Oresme's approach of using diagrams to depict the relationship of two variables.

3.2   Fermat studies the maximums of curves

An amateur mathematician with the mind of a professional, **Pierre Fermat** (France, born 1601) solved some of the toughest problems of his day and even proposed his own to the community. He was attracted to the same sorts of puzzles that occupied Oresme, including the questions of what there is to learn about curves and what tools best assist in this task.

Fermat inherited a tool that Oresme did not possess: the sophisticated, symbolic approach of **François Viète** (France, born 1540). Before Viète, most mathematical exposition relied on verbal descriptions of the subject, and part of Viète's program aimed to encode such explanations in clear, precise notation. For centuries, mathematicians held that a curve resulted from the intersection of two geometric objects; a parabola, for instance, is created when a plane intersects a cone just so. Fermat, augmenting what he had gained from Viète, asserted that a curve also results whenever two unknown quantities enjoy some relationship as expressed by an equation. Thus, a parabola is born at the mere mention of $y = x^2$.

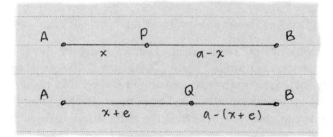

Figure 3.4. There are two solutions if the lengths of the cut segments multiply to a number smaller than the maximum.

Fermat turned this innovation on classic problems of geometry like this one: cut a line segment AB of length a (Figure 3.4, top diagram) into two parts, one of length x and the other of length $a - x$, so as to maximize the product $x(a - x)$ of their lengths. The solution — cut the segment in half — was well-known, but Fermat developed a novel algebraic method for solving it. Suppose we do not wish to maximize the product, but instead want cut the segment at a point so that the product of the two lengths equals a number N that is less than the maximum. Then there should be two different cuts that solve the problem. Say that P provides one such solution and that Q provides the other, as in Figure 3.4. With $AP = x$, let $AQ = x + e$, so that e is the distance between P and Q. Both products equal N, so they are themselves equal, hence

$$x(a - x) = (x + e)(a - (x + e))$$
$$ax - x^2 = ax + ae - x^2 - 2xe - e^2$$
$$0 = ae - 2xe - e^2.$$

Dividing by e yields

$$0 = a - 2x - e. \tag{3.1}$$

Now let's return to the original problem of maximizing the product, where there is a unique solution, as if P and Q become one. In such a case, the difference e is zero, and assigning $e = 0$ in (3.1) gives the well-known solution $x = a/2$.

This concise method applies to problems entirely divorced from any geometric inspiration. Say we want to maximize $ax - x^3$ now. Mimicking what we did above, we replace x with $x + e$ and set the results equal:

$$ax - x^3 = a(x + e) - (x + e)^3$$
$$= ax + ae - x^3 - 3x^2e - 3xe^2 - e^3. \tag{3.2}$$

Collecting like terms and dividing by e yields

$$0 = a - 3x^2 - 3xe - e^2,$$

whereupon we let $e = 0$ to find that $a = 3x^2$.

Primary among the concerns that this otherwise elegant method raises is that of the status of the variable e. Moments after dividing by e we let $e = 0$. Is e alive, or is e a ghost? If it is both, then serious doubt is cast on the logical step that marks its transition from alive to dead.

Aware of this, Fermat took care to justify his method. To take some of the pressure off of the concept of *equals* in the initial step of (3.2), Fermat coined the term *adequals* to use in its place. His use of 'adequals' in an equation signaled that one of the non-zero variables in that equation would later be allowed to equal zero. Further, he argued that when e is nonzero we are in the case of multiple solutions and when $e = 0$ we are simply shifting to the case where there is a unique solution. How persuasive his defenses are to us today is a bit beside the point, because we use a different method now for maximizing (and minimizing) curves, but our modern method bears enough similarity to Fermat's to make him a critical character in the story of calculus.

3.3 *Fermat extends his method to tangent lines*

Even more salient to our story is Fermat's treatment of tangent lines. It is straightforward to define *tangent line* if one has access to modern concepts of function and limit, but we are still several chapters from that vantage point. For our purposes — that is, to show how Fermat investigated tangent lines using his method of finding maximums and minimums — we can use a definition from ancient Greece: a tangent line to a curve intersects the curve at exactly one point.

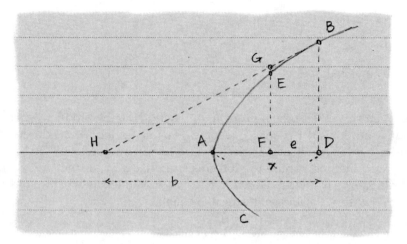

Figure 3.5. Fermat reproduced the well-known result that a tangent line to a parabola intersects the axis so that $b = 2x$.

When Fermat first published his method for tangent lines, he explained how it works using a parabola, but he did not yet realize that such a curve can be expressed simply as $x = y^2$ or $y = x^2$ on coordinate axes. Instead, he used the classical Greek definition of a parabola as that curve BAC, as in Figure 3.5, such that

$$\frac{(BD)^2}{AD} = \frac{(EF)^2}{AF}. \tag{3.3}$$

The vertex A of the parabola lies on the axis HD, oriented horizontally as Fermat drew it. Line BGH is tangent to the curve, so the only point of intersection is B. To construct this tangent line, we must locate H in its proper spot; that is, we must determine the distance from H to A. It was commonly known that A is equidistant from H and D, but Fermat demonstrated this in a new way.

Because G lies outside the parabola on the tangent line, we have $EF < GF$, so (3.3) can reflect this:

$$\frac{(BD)^2}{AD} = \frac{(EF)^2}{AF} < \frac{(GF)^2}{AF}, \quad \text{or} \quad \frac{(BD)^2}{(GF)^2} < \frac{AD}{AF}. \tag{3.4}$$

Now triangles BDH and GFH are similar, so

$$\frac{BD}{GF} = \frac{DH}{FH}, \quad \text{or} \quad \frac{(BD)^2}{(GF)^2} = \frac{(DH)^2}{(FH)^2},$$

which, combined with (3.4), implies that

$$\frac{(DH)^2}{(FH)^2} < \frac{AD}{AF}, \quad \text{or} \quad \frac{(DH)^2}{AD} < \frac{(FH)^2}{AF}. \tag{3.5}$$

If we were to slide F until it were superimposed on D, the left-hand side of this last inequality would remain constant while the right-hand side would decrease and approach that constant, so Fermat had in a sense identified the problem of determining a tangent line with that of finding a minimum value. For this reason, perhaps, he set the two sides of (3.5) adequal to each other; as a reminder, this means that he intends to treat some distance that is currently nonzero as if it is zero. In this case, that distance is FD, which horizontally separates GF from BD.

If we assign $DH = b$, $AD = x$, and $FD = e$, then (3.5) becomes

$$\frac{b^2}{x} < \frac{(b-e)^2}{x-e}, \quad \text{or} \quad \frac{b^2}{x} \approx \frac{(b-e)^2}{x-e}$$

when we let $\approx$ denote that the ratios are adequal. So

$$b^2(x - e) \approx x(b - e)^2 = x(b^2 - 2be + e^2)$$

$$b^2 x - b^2 e \approx b^2 x - 2bxe + xe^2$$

$$b^2 e \approx 2bxe - xe^2.$$

Here Fermat divides by e and then lets $e = 0$ to find that $b^2 = 2bx$ or $b = 2x$, confirming the well-known property of tangent lines to parabolas.

We discussed Fermat's treatment of e earlier, as well as his defense. To be fair, Fermat later developed tools akin to our modern coordinate system and notation, whereupon he addressed this concern of division by zero. But we have studied his original method here not only for the window it opens on the tricky issues that calculus must navigate, but also because we can better discuss an alternative method of finding tangent lines due to someone who might properly be called Fermat's 'rival' in mathematics.

3.4 Descartes proposes a geometric method

Fermat's approach is not the only algebraic way of studying tangent lines. **René Descartes** (France, born 1596) discovered another method, and challenged Fermat to what amounted to mathematical duels using his method as his weapon.

Rather than aim at tangent lines, Descartes showed how to construct *normal lines*, which are perpendicular to tangent lines at the point of tangency. If one line can be found, the other follows easily. Parabola BAC with tangent line BH appears in Figure 3.6 along with normal line BP. While Fermat's method calculates distance AH, Descartes's finds AP.

Descartes argued that any circle with center P will either intersect the upper half of the parabola two times, once, or not at all. If once, then that point must be B. Let r be the length of the radius PB of this circle, and let $AD = x$, $AP = h$, and $BD = a$. Now

$$a^2 = r^2 - (h - x)^2$$

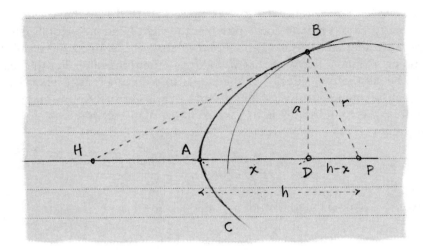

Figure 3.6. The circle touches the curve at one point, so its radius is perpendicular to the tangent line of the curve.

by the Pythagorean Theorem. Using the fact that BAC is a parabola, we know $a^2 = x$, so

$$x = r^2 - (h - x)^2,\tag{3.6}$$

and if the relationship between a and x were different, we would have substituted appropriately.

Now (3.6) may be written

$$x^2 + (1 - 2h)x + (h^2 - r^2) = 0\tag{3.7}$$

as a quadratic equation. Such an equation has two solutions that we can find by factoring the left-hand side. Descartes pointed out that if the circle cut the parabola in two points, then the equation would have two roots; but because the circle only touches the parabola once, the factorization of the left-hand side of (3.7) must be $(x - c)^2$ for some value c. So

$$x^2 + (1 - 2h)x + (h^2 - r^2) = (x - c)^2, \text{ or}$$

$$x^2 + (1 - 2h)x + (h^2 - r^2) = x^2 + (-2c)x + c^2 \,.\tag{3.8}$$

This equation is organized according to decreasing powers of x. The fast way to reach our goal is to *equate coefficients*; that is, the only way for one side of (3.8) to equal the other is if

$$1 - 2h = -2c \quad \text{and} \quad h^2 - r^2 = c^2 \,.\tag{3.9}$$

Descartes stated that when an equation like (3.7) has two equal roots, then in the factoring $(x - c)^2$ the number c equals the unknown quantity x. (For example, the single solution to $(x - 3)^2 = 0$ is $x = 3$.) Thus, we may rewrite (3.9) as

$$1 - 2h = -2x \quad \text{and} \quad h^2 - r^2 = x^2 \,.\tag{3.10}$$

The first of these equalities reveals the relationship $h = x + 1/2$ that we were seeking (and the second equality supports this result when we use the Pythagorean theorem on triangle BDP).

Like Fermat, Descartes could replicate well-known results with an algebraic method that could be applied to curves that did not specifically arise from the study of a geometry problem. But like Fermat, Descartes did not sidestep every logical pitfall. To arrive at (3.9), he expanded $(x - c)^2$ and equated its coefficients to the expression on the left of (3.8). Then to arrive at (3.10), he let $x = c$. But in this case, the expansion of $(x - c)^2$ equals zero, whereupon it becomes problematic to equate coefficients.

And then a *practical* trouble of the method of Descartes lies in the morass of calculations we must navigate; for example, to find (3.6) for a curve such that $a^3 = x$, we land in the unenviable position of having to expand

$$x^2 = \left(r^2 - (h - x)^2\right)^3 \tag{3.11}$$

and then factor $(x - c)^2$ from the resulting sixth-degree polynomial to find the relationship between h and x.

Johannes Hudde (Holland, born 1628) found a clever way to reduce the difficulty of handling equations like (3.11), which helped to make the method of Descartes more accessible. Still, while neither method avoids risky play with the infinite, Fermat's method showed a robustness in meeting new challenges that may explain why modern techniques of studying curves mimic his approach.

3.5 *Furthermore*

3.1 **The infiniteness of the harmonic series.** Mathematicians approach problems the way rock climbers do cliffs: the more difficult the pitch, the more exhilarating the ascent. After a climb has been solved, others look for new routes, or try equipment that no one else has used, simply for the joy of pioneering. The *harmonic series* — so named for its connection to music — illustrates this well.

(a) It seems reasonable that the sum of a series will not change if we group the terms and first add within the groups. We have little reason to doubt, for example, that

$$\left(\frac{1}{2} + \frac{1}{4}\right) + \left(\frac{1}{8} + \frac{1}{16}\right) + \left(\frac{1}{32} + \frac{1}{64}\right) + \cdots = \frac{3}{4} + \frac{3}{16} + \frac{3}{64} + \cdots$$

has a sum other than 1 because of the grouping.

Oresme grouped the terms of the harmonic series

$$1 + \frac{1}{2} + \frac{1}{3} + \frac{1}{4} + \frac{1}{5} + \frac{1}{6} + \frac{1}{7} + \cdots \tag{3.12}$$

in a way that showed that the series is greater than

$$1 + \frac{1}{2} + \frac{1}{2} + \frac{1}{2} + \cdots.$$

Find a grouping that leads to this same conclusion. (And thus, as Oresme put it, the series "becomes infinite." Today, we say that the series *diverges*.)

(b) **Pietro Mengoli** (Italy, born 1626) suggested the grouping

$$1 + \left(\frac{1}{2} + \frac{1}{3} + \frac{1}{4} \right) + \left(\frac{1}{5} + \frac{1}{6} + \frac{1}{7} \right) + \cdots ,$$

arguing that the sum of each group exceeds three times the middle term. Why could he claim this, and how does his claim lead to a proof that the harmonic series becomes infinite?

(Note: yet another proof is offered in exercise 8.1(c).)

3.2 **Oresme compares two series.** Oresme's proof in exercise 3.1 that the harmonic series cannot be summed is justly famous, but he devoted substantially more of his writing to observations on the infinite. Here we look at his reflections on pairs of series that grow at different rates.

Oresme claimed that if we subtract a part of a quantity from itself and repeatedly subtract the same proportion of what remains from the remainders themselves, then the entirety of the quantity will be exactly consumed. Using modern notation, let Q denote the quantity, and suppose that we remove $1/n$ part of Q, leaving

$$Q - \frac{1}{n}Q, \quad \text{or} \quad Q\left(1 - \frac{1}{n} \right).$$

Removing another $1/n$ part of this remainder leaves

$$Q\left(1 - \frac{1}{n} \right) - \frac{1}{n}Q\left(1 - \frac{1}{n} \right), \quad \text{or} \quad Q\left(1 - \frac{1}{n} \right)^2 .$$

Oresme claimed that this process will ultimately deplete Q exactly, or

$$Q = \frac{1}{n}Q + \frac{1}{n}Q\left(1 - \frac{1}{n} \right) + \frac{1}{n}Q\left(1 - \frac{1}{n} \right)^2 + \cdots$$
$$= \frac{1}{n}Q\left[1 + \left(1 - \frac{1}{n} \right) + \left(1 - \frac{1}{n} \right)^2 + \cdots \right]$$

in symbols. What we see in the last set of brackets looks like the series from exercise 2.3, with $x = (1 - 1/n) < 1$. If we can trust that series, then how would it help us establish the truth of Oresme's claim?

Historical note. Although Oresme proved that the series in exercise 2.3 is trustworthy, he also addressed this common-sense objection: Suppose we let $n = 2$ on one hand and $n = 5$ on the other[1]. In the first case, we repeatedly reduce what is left of our quantity by half, whereas in the second case we more slowly

_____
[1]Oresme used $n = 2$ and $n = 1000$, but Figure 3.7 would be difficult to interpret.

reduce what is left by only a fifth part. Pause this process so that each reduction has occurred the same number of steps; what remains of the first quantity is clearly less than what remains in the second, as Figure 3.7 attempts to make clear. So if the second process is always lagging behind the first, then how could both processes exactly consume the same original quantity? This sounds paradoxical.

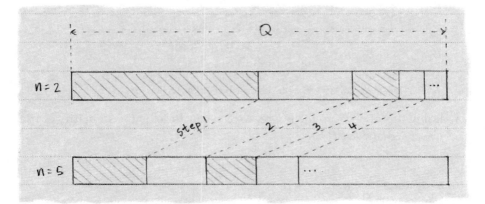

Figure 3.7. This is a geometric representation of a quantity being reduced at two different rates.

Oresme countered this argument by challenging the idea that we may only compare the two remainders after the *same* number of steps. Granted, if we pause at each step and compare what remains of the quantity that is being halved to what remains when $n = 5$, then the first remainder will always be smaller. However, if we freeze this small remainder and continue removing the fifth part of the other, this other remainder will eventually become less than the first. Although the rates differ, then, the quantity cannot escape its fate; it is consumed either way. Oresme's mature approach to series compares favorably with our modern treatment, which appears much later, in section 9.4.

3.3 **Descartes corrects Fermat's method.** In taking issue with Fermat's methods described in this chapter, Descartes wrote a letter (to a third party) in which he hoped to repair the methods. In the letter, Descartes chose as his example the curve ABD (in Figure 3.8) that today we would denote $x = y^3$, although, like Fermat, he defined the curve as those points satisfying

$$\frac{(BC)^3}{AC} = \frac{(DF)^3}{AF}. \tag{3.13}$$

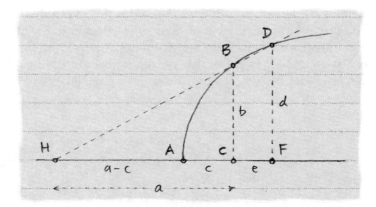

Figure 3.8. Today we would describe curve ABD using $x = y^3$.

(a) Using the labels from the figure, we can conclude by the similarity of triangles ECB and EFD that
$$\frac{b}{d} = \frac{a}{a+e},$$
and from (3.13) that
$$\frac{b^3}{d^3} = \frac{c}{c+e}.$$
Combine these observations and provide the algebra required to show that
$$a^3 = 3a^2c + 3ace + ce^2. \tag{3.14}$$

(b) Line HBD is not tangent to the curve, but if we allow DF to slide left until it is superimposed on BC (which, it is worth noting, would slide H closer to A), then the new line HB would be tangent. Explain how this process transforms (3.14) into $a = 3c$, and then express the significance of this conclusion in your own words.

3.4 **Binomial expansion.** Fermat's approach to finding the slopes of tangent lines and maximum points requires the expansion of expressions like $(x + e)^3$ in (3.2). Such expressions having two terms x and e within the parentheses are called *binomials*. The importance of their expansions will only grow as we proceed with the story of calculus. In this exercise, we set the scene for a thinker whose insight into binomial expansions is detailed in exercise 6.5.

Take a look at the patterns that arise in an expansion where the exponent is an integer. The expansion of $(x + e)^5$ makes the point. By hand, we expand
$$(x + e)^5 = x^5 + 5x^4e + 10x^3e^2 + 10x^2e^3 + 5xe^4 + e^5$$
and observe the pattern $1, 5, 10, 10, 5, 1$ of the coefficients, a signpost to a possible shortcut. To discover the shortcut, we observe that from each factor in
$$(x + e)^5 = (x + e)(x + e)(x + e)(x + e)(x + e)$$

we must choose one term — either x or e — and then find the product of these five choices. To generate x^5 in the expansion, for example, we would choose x from all five factors. Because this is the only way to get the term x^5, its coefficient is simply 1.

The term x^4e requires the choice of x from 4 of the factors and e from the remaining 1 factor. There are 5 factors from which we may choose e, so x^4e appears 5 times in the expansion. Similarly, we get x^3e^2 whenever we choose x from 3 of the factors and e from the other 2 factors. Counting by hand gives 10 ways for this to happen, but counting in this way is tedious.

As with most endeavors, tedium in mathematics points to shortcuts; patterns lead the way there. For starters, why is the number of copies of x^3e^2 the same as the number of copies of x^2e^3 in the expansion? It is because x and e differ only in name. If we choose x from the middle three factors, for example, and therefore e from the first and last to generate x^3e^2, we could just as well have chosen e from the middle three factors to create x^2e^3. Each choice of x from three factors corresponds exactly to one choice of e from the same three factors.

Fair enough; now, why are there 10 such ways to choose three of the five factors? Reframing the question may bring it into sharper focus. Instead of factors, imagine 5 poker chips. How many ways can we choose 3 of them? If we label the chips A, B, C, D, E, then we have 5 choices for the first chip, 4 choices for the second, and 3 choices for our third. Say we choose D, E, A in that order. The number of such choices, then, totals $5 \times 4 \times 3 = 60$. The choice DEA is one of these.

But so also are the other five versions of DEA, namely ADE, AED, DAE, EAD, and EDA. We want to count all 6 of these results as just one, because the chips only differ in their labels. We can take care of such repeats by dividing 60 by 6. Thus, we may choose 3 of 5 identical objects in 10 ways.

What formula does the argument suggest? Starting with n identical objects, we temporarily label them. To choose k of the objects, we first choose from the original n, then the remaining $n - 1$, then the remaining $n - 2$, and so on, until we have chosen k times. But for each such choice, there are several others that will repeat that choice once we remove the labels. Specifically, we need to divide by the number of ways that we can rearrange those k labeled objects, which is

$$k \times (k - 1) \times (k - 2) \times \cdots \times 1 \, .$$

Mathematicians denote this product with the symbol $k!$ or the language 'k factorial'.

One last tricky bit before we can generate the formula: when we choose k of n labeled objects, there are $n - k$ objects remaining, so our kth and final choice came from $n - k + 1$ objects. Hence, the number of ways to choose k of n identical objects is

$$\frac{n(n-1)(n-2)\cdots(n-k+1)}{k(k-1)(k-2)\cdots 1}\,. \tag{3.15}$$

Multiplying both numerator and denominator by $(n-k)!$ yields the modified and better-known version

$$\binom{n}{k} = \frac{n!}{k!(n-k)!} \tag{3.16}$$

where the left-hand side is read 'n choose k' and the right-hand side is a simplified version of (3.15). So, for example, we might write

$$\binom{6}{3} = \frac{6!}{3!\,3!} = \frac{6\cdot 5\cdot 4\cdot 3\cdot 2\cdot 1}{3\cdot 2\cdot 1\cdot 3\cdot 2\cdot 1} = 20$$

to find the number of ways of choosing 3 of 6 identical objects.

(a) List these 20 possibilities, using the illustration of poker chips labeled A, B, C, D, E, F.

(b) Expand $(x+e)^6$ using the techniques just discussed, rather than expanding by hand.

(c) Using summation notation and the notation of (3.16), write a sum that equals $(x+e)^n$ for a positive integer n. This equality is called the *binomial theorem*.

4

Indivisibles

Every mathematical subject advances thanks to imaginative conjectures. One of the earliest examples of such risk-taking in calculus is due to **Democritus** (Greece, born *c.* 460 BCE), who lived about 200 years before Archimedes. He is credited with a claim such as the following:

> If two solids are cut by a plane parallel to their bases and at equal distances to their bases, and the sections cut by the plane are equal, and if this is true for all such planes, then the two solids have equal volumes.

Although this claim does not directly state that solids are composed of infinitely many two-dimensional slices, it certainly toys with the idea. One might ask, for example, what becomes of the topmost slice of a pyramid, at its tip. Do we jump from two dimensions to only one? Is the jump sudden, or gradual? In fact, Democritus himself skeptically inquired if two infinitely thin slices of a solid could be neighbors. Yet despite puzzles like this, mathematicians used the statement above to compare the volumes of cylinders, cones, prisms, pyramids.

Inspired leaps leading to truth — it is no wonder that some have claimed that revealing truths in mathematics takes as much creativity as in the arts and letters. In this chapter, we see how European mathematicians engaged in this pursuit.

4.1 Cavalieri's quadrature of the parabola

The simplest area to calculate is that of a square, and because the prefix *quad-* suggests squares, we use the phrase 'a *quadrature* of' to mean 'an area equal to'. We use this term even when the area is not square, as when Archimedes found a quadrature of a parabolic segment.

In 1639, **Bonaventura Cavalieri** (Italy, born 1598) published a method for finding the area bounded by a parabola; his argument echoes the statement attributed to Democritus above. Cavalieri's quadrature of the parabola, in fact, began with a square as in Figure 4.1. Segment EF bisects square $ABCD$ and meets diagonal

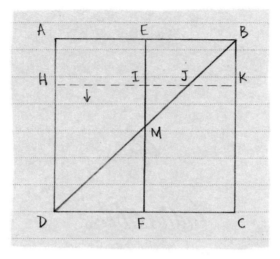

Figure 4.1. Segment *HK* remains parallel to *AB* as it slides through the square *ABCD*.

BD at *M*. Dotted segment *HK* remains parallel to *AB* as it slides from *AB* down to *DC*. It meets *EF* at *I* and *BD* at *J*.

Upon *HJ*, Cavalieri imagined building a square. In Figure 4.2, we see *ABCD* lying 'flat' with shaded square *HJPQ* built upon *HJ*. As *HK* slides from *AB* to *CD*, the square on *HJ* decreases in area until, ultimately, it vanishes to a point at *D*. The volume created by this sliding square resembles a four-sided pyramid lying on its side, with square base *ABRS* and tip *D*. For convenience, we let △ denote this pyramid.

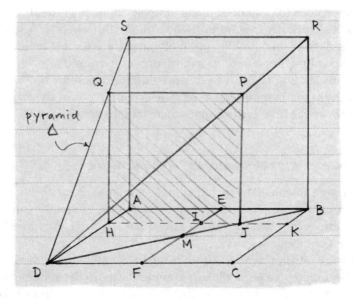

Figure 4.2. Imagine the square *HJPQ*, built on *HJ*, sweeping through pyramid △.

Cavalieri argued that the volume of $\triangle$ equals the area DAL under the parabola in Figure 4.3. Here is why: segment HN is the square of DH, and $DH = HJ$ in Figure 4.2, where the 'square of HJ' is shaded; in other words, the length of HN in Figure 4.3 is the area of square $HJPQ$ in Figure 4.2. Thus, Cavalieri claimed that the *area* created as HN sweeps from AL to D is the *volume* created as $HJPQ$ sweeps from $ABRS$ to D.

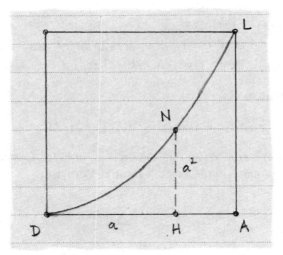

Figure 4.3. The area of region DAL equals the volume of pyramid $\triangle$ in Figure 4.2.

A reader blessed with good spatial relations can see that pyramid $\triangle$ in Figure 4.2 occupies one-third of the cube that would form if $ABRS$ swept parallel to itself until it stood on CD. (The frontispiece shows that three copies of $\triangle$ assemble into that cube.) Not wishing to rely on such visual or physical observations, however, Cavalieri expressed his argument using words and algebra. When he wrote "all squares on HJ", for example, he means the volume created by all possible squares built on HJ as it sweeps from AB to D; that is, he meant the volume of $\triangle$.

We can more easily follow his argument if we replace phrases like "all squares on HJ" with notation. Using summation notation, we let

$$\sum_{A}^{D} (HJ)^2$$

denote "all squares on HJ", where A and D indicate the range through which the squares on HJ sweep. Our goal, stated with this notation, is to show that

$$\sum_{A}^{D} (HJ)^2 = \frac{1}{3} \sum_{A}^{D} (AB)^2 . \tag{4.1}$$

Cavalieri began with an observation about Figure 4.2:

$$(AB)^2 = (HJ + JK)^2 = (HJ)^2 + 2(HJ \cdot JK) + (JK)^2 . \tag{4.2}$$

Cavalieri focused on the product $HJ \cdot JK$, observing that

$$IJ = HJ - HI$$
$$\implies IJ = HJ - IK, \text{ and}$$
$$IJ = IK - JK.$$

Adding the second and third equations yields

$$2(IJ) = HJ - JK$$
$$\implies 4(IJ)^2 = (HJ - JK)^2 = (HJ)^2 - 2(HJ \cdot JK) + (JK)^2$$
$$\implies 2(HJ \cdot JK) = (HJ)^2 + (JK)^2 - 4(IJ)^2 . \tag{4.3}$$

Using (4.3) in (4.2) gives

$$(AB)^2 = 2(HJ)^2 + 2(JK)^2 - 4(IJ)^2 . \tag{4.4}$$

Now because (4.4) is true for any particular segment HK, Cavalieri claimed that the collection of *all* squares will obey the same pattern; thus,

$$\sum_A^D (AB)^2 = 2\sum_A^D (HJ)^2 + 2\sum_B^C (JK)^2 - 4\sum_E^F (IJ)^2 . \tag{4.5}$$

Visualizing in three dimensions (as we did in Figure 4.2) supports Cavalieri's observation that

$$\sum_A^D (HJ)^2 = \sum_B^C (JK)^2 , \tag{4.6}$$

as the two volumes denoted by these sums are mirror images. Similarly,

$$\sum_E^F (IJ)^2 = 2\sum_E^M (IJ)^2 . \tag{4.7}$$

Finally, Cavalieri imaginatively suggested that we picture the squares on IJ sweeping from E to M during the same *time* as it takes the squares on HJ to sweep from A to D. Because EM is half of AD, the squares on HJ will have to move twice as fast — so there are twice as *many* — compared to the squares on IJ. Further, each segment IJ corresponds to a segment HJ that is twice as long, so the *square* on HJ will have *four* times the area of the square on IJ. The upshot of these thoughts is that

$$\sum_A^D (HJ)^2 = 8\sum_E^M (IJ)^2 . \tag{4.8}$$

Combining (4.8) with (4.7) yields

$$4\sum_E^F (IJ)^2 = \sum_A^D (HJ)^2 ,$$

which, along with (4.6), allows us to substitute into (4.5):

$$\sum_A^D (AB)^2 = 2\sum_A^D (HJ)^2 + 2\sum_A^D (HJ)^2 - \sum_A^D (HJ)^2$$

$$= 3\sum_A^D (HJ)^2 \ .$$

Dividing by 3 produces equation (4.1) and its implications for the area under the parabola.

Cavalieri used the word *indivisibles* to describe objects like the "squares on HJ" that cannot be sliced any thinner. His idea that solids were composed of indivisibles correctly solved the problem we just studied and many more besides; yet in its wake, puzzling questions remain. As one example, look at the paragraph preceding (4.8). Cavalieri claimed that there are twice as many squares on HJ from A to D as there are squares on IJ from E to M, yet he then compared the areas of *one* square from each group. Does this leave out half of the squares on HJ? Does it help that both groups of squares are infinite?

Cavalieri cared about the answers to these questions, but lack of answers did not stop him from advancing similar arguments. In fact, he went on to compute the areas under curves of degree higher than two, ultimately asserting that

$$\frac{\text{area under the curve } y = x^k}{\text{between the vertex and } x = a} = \frac{a^{k+1}}{k+1} . \tag{4.9}$$

Despite the puzzling nature of his arguments, his results were correct; better yet, they were crucial building blocks in the story of calculus.

4.2 *Roberval's quadrature of the cycloid*

An admirer of Cavalieri's methods, **Gilles Personne de Roberval** (France, born 1602) found brilliant uses for them. One of Roberval's quadratures illustrates this particularly well.

If the circle with diameter AB in Figure 4.4 were to roll to the right while sitting on AG, the point A would track along the dashed curve, peaking at point C, and returning to the 'ground' at point G. The curve ACG is called a *cycloid* because it is generated by a rolling circle; the point A acts like a tack stuck in a bicycle tire. The cycloid caught the fancy of mathematicians in the late 1600s in part because it provided the surprising solution to questions like these two: Along what curve would a clock's pendulum swing so that the clock would keep perfect time no matter how far the pendulum traveled in one swing? What curve allows a ball placed on it to descend from one point to another in the fastest time? In each case, an inverted version of the cycloid shown in Figure 4.4 answers the question.

Galileo, teacher of Cavalieri, attempted a quadrature of the cycloid by building one of metal and weighing it against the metal circle that generated it. This experiment suggested to Galileo that the cycloid's quadrature was about three times that of the generating circle. Roberval proved that it was *exactly* three times.

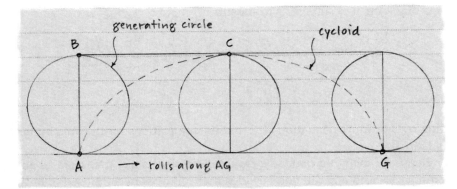

Figure 4.4. The cycloid ACG tracks the movement of point A on the circle as it rolls from A to G.

Roberval's argument is easier to follow if we focus on the latter half of the circle's journey as it is depicted in Figure 4.4. In Figure 4.5 (a), focus on point C as it moves to C', tracing part of a cycloid. (Points P and D move to P' and D', respectively, each one tracing a part of its own cycloid, too.)

When the circle begins to roll, imagine that the original circle remains behind, motionless, like a ghost of the rolling circle — see Figure 4.5 (b). Let L be the point on CP that mirrors the height of C', so LC' is always parallel to PG. Points N and M lie on LC' so that N remains on the ghost circle and M stays directly above the rolling circle's point of contact with the ground. Roberval noted that $LN = MC'$ at all times; as LN sweeps through the semicircle CNP, segment MC' sweeps through part of the region under the cycloid $CC'G$. Thus, the area of the semicircle CNP equals the area through which MC' sweeps.

Thus, the path that M takes cuts the desired area under the cylcoid into two unequal parts, one of which we have identified with half of the area of the rolling circle. We can find the area of the other part if we come to understand the details of how M moves.

Suppose that the circle rolls at a constant speed; then N travels on the ghost circle from C to P at a constant speed. Because M stays directly above the rolling circle's point of contact with PG, its horizontal speed is constant. However, its vertical speed mirrors that of N, which moves downward fastest when it is directly to the right of the center of the ghost circle. Although the vertical speeds of N and M are changing, the speeds possess symmetry, leading Roberval to conclude that the path of M cuts rectangle $CPGH$ in half in the manner shown in Figure 4.5 (c).

The area of $CPGH$ is its height (the diameter of the circle) times its width (half of the circumference of the circle). Thus, if we let r denote the radius of the circle, we find that

$$\text{area swept by } LM = \frac{1}{2}(\text{area of } CPGH)$$
$$= \frac{1}{2}(2r)\frac{1}{2}(2\pi r)$$
$$= \pi r^2$$
$$= \text{area of the rolling circle.}$$

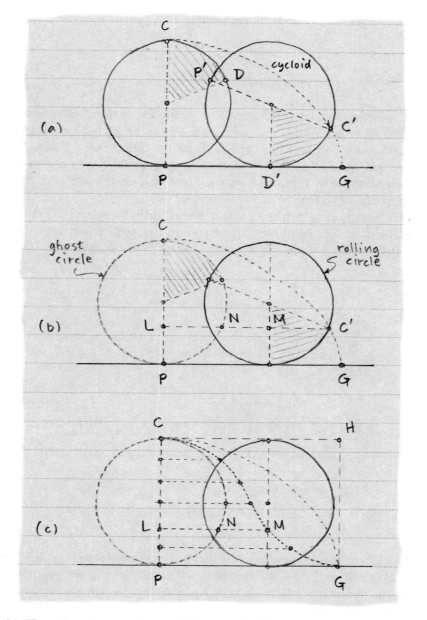

Figure 4.5. These three figures help establish the truth of Roberval's claim about the cycloid.

Combining this observation with the earlier result that MC' sweeps through an area equal to half the area of the rolling circle, Roberval concluded that the area of $CC'GP$ was one and a half times the area of the rolling circle. So Galileo came close when he weighed metal cycloids: the area under a complete cycloid is exactly three times the area of the circle that generates it.

4.3   Worry over indivisibles

Roberval's argument is appealing and its conclusion is true, both hallmarks of excellent mathematics. At its heart, though, lies an assumption like that attributed to Democritus at the start of this chapter: Roberval claims that LN and MC' are equal at all heights as they sweep through semicircle PNC and region $CMGC$ respectively, so these two regions have the same area. On the face of it, this claim seems true. In fact, it *is* true. Why not believe it?

Cavalieri, who did as much as anyone to reveal truths based on this assumption, was challenged on this point by mathematicians like **Paul Guldin** (Switzerland, born 1577). Consider the triangle in Figure 4.6, for example. Altitude BD divides

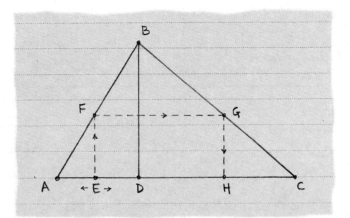

Figure 4.6. This figure challenges the idea that areas are composed of indivisible lines.

triangle ABC into two regions. Imagine that E moves from A to D with EF parallel to BD. At all times, let FG be parallel to AC with G on side BC, and H on AC so that GH is parallel to EF. No matter where E sits on AD, the corresponding point H sits on DC, and $EF = HG$. By the assumption used by Roberval in his quadrature, regions ABD and BDC ought to be equal. Clearly they are not.

This challenge troubled Cavalieri, who suggested that there are *more* indivisibles in one region than the other. But how can one infinitude exceed another? Why does Roberval's argument lead to truth while this one does not? Puzzles upon puzzles, exactly what mathematicians love.

4.4 Furthermore

4.1 **Kepler determines the volume of an 'apple'.** A few decades before Cavalieri produced the work described in this chapter, the astronomer **Johannes Kepler** (Germany, born 1571) used indivisibles to calculate the volumes of a wide variety of solids. One such volume was that of a circle that is revolved around one of the points on its perimeter, as in Figure 4.7. Imagine the circle with center Q sweeping through space while hinged at point P on its perimeter; then Q will pass through Q' and a typical chord AB that is parallel to the tangent at P will pass through $A'B'$. The resulting solid looks like a doughnut (or *horn toroid*, to use the mathematical term) with its hole closed.

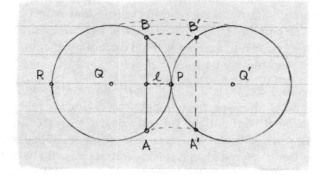

Figure 4.7. Revolving a circle around one of its perimeter points P creates a solid (shown in cross-section here).

Kepler reasoned that, as it revolves, each segment like AB creates a cylinder with height AB and radius equal to the perpendicular to AB through P (this is ℓ in the figure). He imagined unrolling the cylinder to create a rectangle, which he situated on AB perpendicular to the original circle.

As we consider the infinitude of segments like AB that lie between points P and R, the resulting cylinders appear to occupy the whole of the toroid. Kepler claimed that unrolling all of the rectangles creates a new solid with a volume that is easy to calculate.

(a) Imagine that AB sweeps through the circle from P to R at a constant rate. Why do the heights of the unrolled cylinders increase also at a constant rate?

(b) Kepler therefore claimed that the solid created by the unrolled cylinders comprises a cylinder, sliced diagonally in half, having circle Q as its base. What is the height of this cylinder?

(c) Thus, what is the volume of half of this cylinder? (Because this volume equals the volume of the toroid, Kepler has reached his goal.)

Historical note. Revolving circle Q around one of its chords, such as AB in Figure 4.7, we arrive at a solid that resembles an apple. Kepler investigated this and many other related shapes. His use of indivisibles was less rigorous than Cavalieri's, but his work inspired Cavalieri and others thanks in part to his creative and prolific output.

4.2 Torricelli discovers a marvelous shape. The prolific **Evangelista Torricelli** (Italy, born 1608) was acquainted with Galileo and Cavalieri. Like those thinkers, he used indivisibles to discover beautiful results. He particularly celebrated his discovery of a shape that seems to contain infinite volume, but does not.

In his elegant proof, he perceives a three-dimensional region as a composite of infinitely many two-dimensional surfaces. The region, resembling a trumpet with an infinitely long neck, lies within a curve that is revolved around the vertical axis. The curve is a *hyperbola*, shown as LDJ in Figure 4.8; the vertical height of each point of a hyperbola is the reciprocal of the horizontal distance from A; so, for example, we have $LI = 1/AI$.

Choose any point C to the right of A along the horizontal and let D be the corresponding point on the hyperbola. Imagine revolving the region $BACDL$ (despite its infinite height) around the vertical axis AB; then DC will pass through EF and L will pass through N to create the trumpet-shaped solid that Torricelli wishes to study. At first glance, it is clear that this solid (which we will call T) has infinite volume. Torricelli enthusiastically proved this wrong.

First, we must view the volume of T through Torricelli's eyes, as an infinite collection of cylindrical surfaces. Each segment like LI creates the curved surface of a cylinder as it revolves around the vertical axis. The (shaded) rectangle $OILN$ is a side-view of this particular cylinder. As we sweep from DC to BA, each segment like LI creates such a curved surface. These indivisibles occupy the full volume of T. (As the cylinders constrict ever closer to AB, matters become strange. We overlook this for now.)

Torricelli concocts a new surface that has the same area as each indivisible cylindrical surface such as the one created by LI. The surface area of the shaded cylinder, viewed as an uncurled rectangle, equals its height LI times its length, which is the circumference $2\pi(AI)$ of the cylinder's base. Because $LI = 1/AI$, the surface area of the shaded cylinder is simply 2π. This is the area of a circle with diameter $2\sqrt{2}$, so we construct rectangle $AHGC$ so that $AH = 2\sqrt{2}$.

View $AHGC$ as a side view of a cylinder revolved vertically around its central axis, labeled ℓ in the figure. Extend LI through I to M, and see IM as slicing the new cylinder. The resulting circle has area 2π.

(a) In your own words, explain how the figure appears to justify Torricelli's claim that he had discovered an infinitely long shape that contained a finite volume.

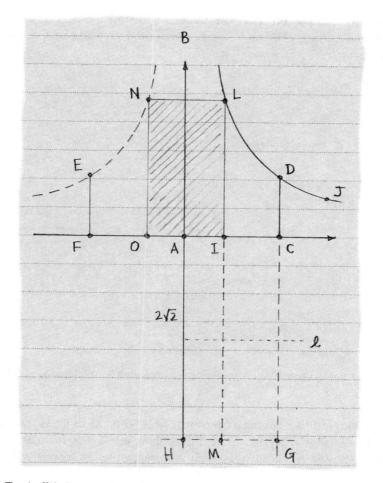

Figure 4.8. Torricelli's 'trumpet' results from revolving region $BACDL$ around the vertical axis.

(b) What happens as we consider segments like LI that approach the vertical axis AB? Does this consideration impact Torricelli's argument, in your opinion?

Historical note. Aware that indivisibles imply perplexing beliefs about the infinite, Torricelli also presented a proof in a style considered more rigorous by many mathematicians of the day; more on this in exercise 6.4. Where arguments using indivisibles lack rigor, however, they often offer brevity and clarity.

4.3 **Mei and Valerio investigate the volume of a sphere.** The manipulation of a sphere by **Mei Wending** (China, born 1633) attests to his mastery of the technique of indivisibles. Think of the volume of a sphere as the collection of infinitely many indivisible circles like the one shaded in Figure 4.9. Its center is

M and any point P on its circumference is distance r from the center O of the sphere. Letting $OM = h$, we can write the area of the shaded circle as $\pi(MP)^2 = \pi(r^2 - h^2) = \pi r^2 - \pi h^2$.

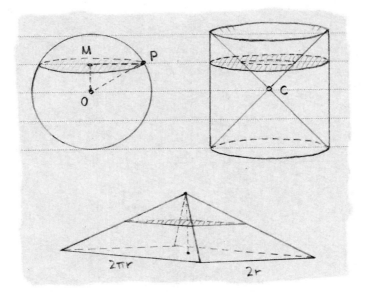

Figure 4.9. Mei transformed a solid sphere first into a 'deficient' cylinder, and then into a pyramid.

A circle with radius r having a concentric hole with radius h has this same area. Using these 'rings', Mei refashioned the sphere into a new form, the 'deficient' cylinder, shown in the figure, which is missing the two cones with tips that meet at C.

(a) Explain why the cylinder has height $2r$.

(b) Another clever transformation carries the deficient cylinder to a pyramid with rectangular base measuring $2\pi r$ by $2r$ and with height r. It is not the shaded ring in the deficient cylinder, however, that shares its area with the shaded rectangle in the pyramid, but rather a different shape within the deficient cylinder. What is it?

Historical note. Mei used these transformations to establish that the volume of a sphere is two-thirds that of the smallest cylinder that encloses the sphere. This result was already well-known, but in this case, it was his method that is notable.

(c) In a similar vein, **Luca Valerio** (Italy, born 1552) invented a single figure that he used to reveal the same truth as did Mei. In Figure 4.10, we see a side-view of a hemisphere AFB with center C and diameter AB, a cone

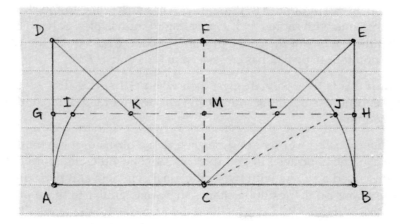

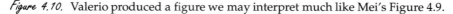

Figure 4.10. Valerio produced a figure we may interpret much like Mei's Figure 4.9.

with tip at C and base DE, and a cylinder sharing that base with the cone and its other base with the hemisphere. Let GH represent any plane cutting these shapes; it cuts the side-views of the hemisphere at points I and J and the cone at points K and L. Point M lies at the intersection of GH and the central axis CF of the cylinder.

By introducing the (dotted) segment CJ and using the Pythagorean Theorem, Valerio established that

$$ML^2 = CB^2 - MJ^2 . \qquad (4.10)$$

Explain the details of this work.

(d) Dividing by CB^2, Valerio expressed (4.10) as

$$\frac{ML^2}{FE^2} = 1 - \frac{MJ^2}{CB^2} . \qquad (4.11)$$

Interpret each fraction in (4.11) as the ratios of the areas of the circles having the relevant segments as radii. Then consider the implications of these ratios as GH sweeps from DE to AB. Assuming Mei's conclusion in the 'Historical note' in exercise 4.3(b), what relationship does (4.11) suggest about the volumes of cones and cylinders?

Historical note. Valerio did not in fact pursue quite the line of reasoning suggested here, but rather considered thin cylinders using the circles as bases.In this, he joins ibn al-Haytham among those thinkers who presaged our modern view of these matters.

(e) Galileo (via Salviati, a character in his book *Dialogue Concerning the Two Chief World Systems*) highlighted a paradox hidden in Figure 4.10. Using Valerio's equation (4.10), Salviati argues that the circular ribbon created by revolving GI around FC through HJ has area equal to the circular base of the cone with radius ML. Appealing to indivisibles, Salviati concludes that the volume of the cone with cross-section CKL equals the volume of the pointed shape that has cross-section GIA and HJB.

But if this is true, Salviati says, then when GH slides all the way to AB, we encounter something strange. In your own words, what paradox do we encounter?

Historical note. Over ten centuries before the births of these thinkers, Chinese mathematicians **Liu Hui** (born c. 230) and **Zu Geng** (born c. 450) circumscribed a sphere with two orthogonal cylinders and performed transformations on the resulting shapes that were similar to those undertaken by Mei Wending. Zu Geng justified the process with the words, "Since volumes are made up of piled-up blocks, [it follows that] if the corresponding areas are equal then their volumes cannot be unequal."

5

Quadrature

Three hundred fifty cities in the world
Just thirty teeth inside of our heads
These are the limits to our experience
It's scary, but it's all right
And everything is finite.

— from the album *Feelings*, David Byrne (1997)

Amid a flurry of discovery in the 1600s, European mathematicians began to recognize the underlying unity of their results. Quite often, a successful quadrature shed light on the mysteries inherent in series, or a clever use of geometry prompted advances in the theory of numbers. Each new connection fanned the intellectual fire. We see this effect in this chapter as we trace efforts to find the quadrature of the hyperbola.

5.1 Gregory studies hyperbolas

One way to define a hyperbola is as the collection of points whose horizontal and vertical components are inversely proportional, as in the modern[1] notation $xy = 1$ or $y = 1/x$. Figure 5.1 depicts one of the two branches of a hyperbola. A successful quadrature of this curve would answer any question like, 'What is the area labeled A in the figure?'

Gregory of Saint-Vincent (Belgium, born 1584) took the first step, finding an *arithmetic sequence* linked to a *geometric sequence* in Figure 5.1. A sequence of numbers is *arithmetic* if each member equals the previous member plus some common constant (described in more detail in exercise 2.1). In contrast, a sequence is *geometric* if each member equals the previous member *times* some common constant, as in the sequences

$$2, 6, 18, 54, \ldots \text{ and } 1, 1/3, 1/9, 1/27, \ldots .$$

---

[1] At this point, we will begin to use the notation for curves and the coordinate system of modern day, simply to ease reading, despite being decades away (in the story) from their full maturation.

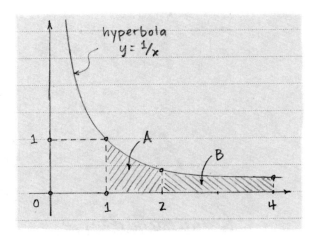

Figure 5.1. Because 1, 2, 4 is a geometric sequence, areas A and B are equal.

Gregory discovered that if we draw vertical lines along the horizontal at distances that are in a geometric sequence, then the areas under the hyperbola within those vertical lines are all equal. For example, because 1, 2, 4 is a geometric sequence, Gregory found that areas A and B are equal in Figure 5.1.

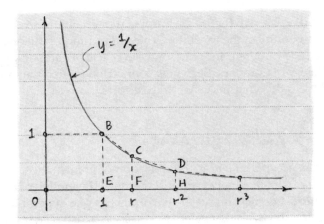

Figure 5.2. Gregory of Saint-Vincent approximated the area under the hyperbola with trapezoids.

The constant multiplied by each member of a geometric sequence to get the next is called the *ratio*, so we will let r stand for this ratio and begin our sequence at 1 to get $1, r, r^2, r^3, \ldots$ as seen along the horizontal in Figure 5.2. (The case in the figure represents $r > 1$.) Gregory argued that trapezoids $BCFE$ and $CDHF$ are equal in area; he then subdivided each trapezoid and argued the same for the four new trapezoids, and so on. He assumed that as the trapezoids become more numerous, the areas of error (where each trapezoid extends up past the hyperbola) would vanish.

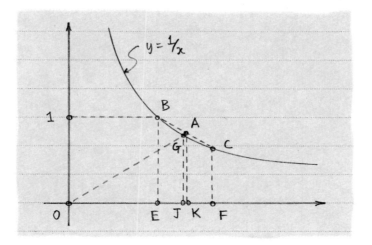

Figure 5.3. The point J subdivides the horizontal into a new geometric sequence.

That trapezoids $BCFE$ and $CDHF$ have equal area follows from

$$\text{area of } BCFE = \frac{1}{2}(BE + CF)(EF) = \frac{1}{2}\left(1 + \frac{1}{r}\right)(r - 1),$$

$$\text{area of } CDHF = \frac{1}{2}(CF + DH)(FH) = \frac{1}{2}\left(\frac{1}{r} + \frac{1}{r^2}\right)(r^2 - r).$$

Now we turn to the subdivision of $BCFE$. Locate the midpoint A of BC and let G be the point where OA intersects the hyperbola, as in Figure 5.3. Then A has coordinates $((1/2)(1 + r), (1/2)(1 + 1/r))$ while G is simply assigned the coordinates $(g, 1/g)$, where $g = GJ$. By the similarity of triangles OGJ and OAK we may write

$$\frac{g}{(1/2)(1 + r)} = \frac{1/g}{(1/2)(1 + 1/r)} \implies \frac{1}{g} = \frac{g}{r} \, ,$$

or $g = \sqrt{r}$. The sequence $1, g, r$ is therefore geometric with common ratio $\sqrt{r}$, so trapezoids $BGJE$ and $GCFJ$ will have equal area just as the original pair did. Nothing prevents us from extending this argument to trapezoid $CDHF$, and then to other initial pairs of trapezoids. Gregory assumed that further subdividing would ultimately exhaust the desired region under the hyperbola with no error.

5.2   De Sarasa invokes logarithms

Gregory, a member of the Jesuit order, studied with an admirer of Galileo. **Alphonse Antonio de Sarasa** (Spain, born 1618) became Gregory's student after joining the Jesuits at age fourteen. Later, the two were colleagues on a college faculty, during which time de Sarasa linked Gregory's result on hyperbolas to recent work by other European mathematicians.

These others (primarily **John Napier** in Scotland and **Henry Briggs** in England) had discovered a way to transform multiplication problems into addition problems (and division into subtraction). All they required was a matching between a

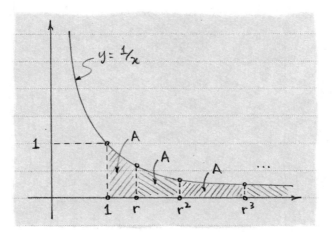

Figure 5.4. The areas described by Gregory of Saint-Vincent are logarithms of the sequence along the horizontal.

geometric sequence and an arithmetic sequence, like the one in Table 1. To multiply 4 and 8, for example, we find their corresponding entries 2 and 3 (in the second column), then *add* 2 and 3, and finally look at the match for the sum (back in the first column). The answer, therefore, is 32.

geometric sequence	arithmetic sequence
1	0
2	1
4	2
8	3
16	4
32	5
64	6

Table 1. A logarithmic table.

Each number in the arithmetic sequence is called the *logarithm* of its match in the geometric sequence. Using the common abbreviation, we have $\log 4 = 2, \log 8 = 3$, and so on. If you are familiar with logarithms, you automatically ask what the *base* is in this example; but the thinkers who invented logarithms did not ask this question. It was enough just to have a matching between two sequences like these.

Further, you might wonder how this matching would help us multiply, say, 3 and 5. It would not; we would require a more complete geometric sequence. The inventors generated these, to varying degrees of precision; Briggs, for example, provided a table of logarithms accurate to over ten decimal places. (His approach is outlined in exercise 5.4, where the term *base* is defined.)

The important feature of logarithms for de Sarasa was the matching of the sequences, one arithmetic and one geometric, and Gregory's work provided just that. As we have seen, Gregory showed that if we mark off a geometric sequence $1, r, r^2, r^3, \ldots$ along one axis of a hyperbola, then the subdivided areas are all equal. In Figure 5.4, these areas are all denoted A.

marks along horizontal	total area from $x = 1$ to this mark
1	0
r	A
r^2	$2A$
r^3	$3A$
r^4	$4A$
$\vdots$	$\vdots$

Table 2. The logarithmic matching found in Figure 5.4.

As Table 2 shows, we therefore have a geometric sequence matched with an arithmetic sequence. Thus we may say that the logarithm of each horizontal value is related to the corresponding area under the hyperbola. Because r may be any ratio we desire, we conclude that for any $a \geq 1$,

$$\log a = \quad \text{the area under the hyperbola } y = 1/x$$
$$\text{between } x = 1 \text{ and } x = a.$$

In a bit, we shall shift our perspective slightly by moving the hyperbola one unit to the left so that it crosses the vertical axis one unit above the origin; in modern notation, the equation of this hyperbola is $y = 1/(1 + x)$. Then

$$\log(1 + a) = \quad \text{the area under the hyperbola } y = 1/x$$
$$\text{between } x = 1 \text{ and } x = 1 + a$$

$$= \quad \text{the area under the hyperbola } y = 1/(1 + x) \qquad (5.1)$$
$$\text{between } x = 0 \text{ and } x = a.$$

5.3 *Brouncker finds a quadrature of a hyperbola*

Gregory's work opened the door for de Sarasa to connect hyperbolas and logarithms, but neither thinker produced an actual quadrature of the hyperbola. A quadrature was found in about 1655 by **William Brouncker** (Ireland, born 1620) thanks to a clever covering of the desired area by rectangles. He focused on the region between $x = 1$ and $x = 2$ under the hyperbola; in Figure 5.5, this is the region between AE and BC.

Cutting rectangle $ABCD$ (having area $1/2$) from the region leaves us with the unknown area of region DCE. Brouncker took the point G halfway between A and

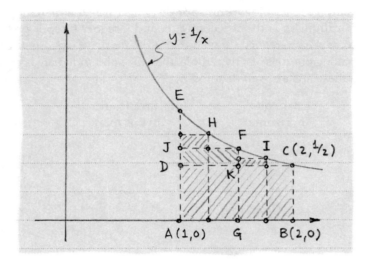

Figure 5.5. Brouncker exhausted the desired area with stacked rectangles.

B and located the corresponding point F $(3/2, 2/3)$ on the curve. Rectangle $DKFJ$ has area $1/2 \cdot (2/3 - 1/2) = 1/12$. Points H $(5/4, 4/5)$ and I $(7/4, 4/7)$ are found by halving segments AG and GB, and the two rectangles with opposite corners JH and KI have respective areas $1/4 \cdot (4/5 - 2/3) = 1/30$ and $1/4 \cdot (4/7 - 1/2) = 1/56$.

Thus far, Brouncker has begun to exhaust the desired area with four rectangles having combined area

$$\frac{1}{2} + \frac{1}{12} + \frac{1}{30} + \frac{1}{56} .$$

Noting a pattern in the denominators, Brouncker conjectured that ultimately the desired area is

$$\frac{1}{1 \cdot 2} + \frac{1}{3 \cdot 4} + \frac{1}{5 \cdot 6} + \frac{1}{7 \cdot 8} + \cdots . \tag{5.2}$$

None of the rectangles extend outside the desired area, so Brouncker might have felt confident that he had successfully quantified the area under the hyperbola; but because his answer was an infinite sum, he treated it with care. We will not look at this part of his argument, but will simply observe that his efforts in this regard were pioneering.

Before we move on, note that (5.2) can be written

$$\frac{1}{1 \cdot 2} + \frac{1}{3 \cdot 4} + \frac{1}{5 \cdot 6} + \cdots = \left(1 - \frac{1}{2}\right) + \left(\frac{1}{3} - \frac{1}{4}\right) + \left(\frac{1}{5} - \frac{1}{6}\right) + \cdots$$

$$= 1 - \frac{1}{2} + \frac{1}{3} - \frac{1}{4} + \frac{1}{5} - \frac{1}{6} + \cdots . \tag{5.3}$$

This series, apart from the alternating signs, mimics (1.5), the series Mengoli proved did not have a finite sum (see exercise 3.1). Alternating the signs is evidently enough for the sum in (5.3) to exist, for it equals the area Brouncker set out to find.

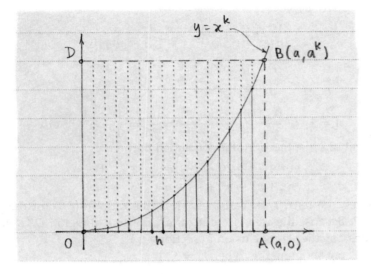

Figure 5.6. Wallis used indivisibles to find his quadrature of the family of curves $y = x^k$.

5.4 Mercator and Wallis finish the task

A combined effort by **Nicholas Mercator** (France, born *c.* 1620) and **John Wallis** (England, born 1616) further unlocked the quadrature of the hyperbola. As a consequence of their discovery, they provided a sum for (5.3). The argument as presented here mimics that in a letter written by Wallis that simplified Mercator's original proof; it relies on a result by Wallis that we turn to first.

As Fermat also had done, Wallis calculated the area beneath curves of the form $y = x^k$ where the integer k may exceed 2. Each such curve has the characteristic shape shown in Figure 5.6. Appealing to the notion of indivisibles, Wallis sliced region OAB into vertical line segments as shown in the figure. If h is the common distance between the n slices, and $OA = a$, then $a = nh$.

As we increase the number of slices (let n grow larger), the slices crowd more closely together (h becomes smaller). Ultimately, the one-dimensional slices will 'fill' region OAB. Wallis expressed the area of region OAB versus the area of rectangle $OABD$ as a ratio:

$$\frac{\text{area } OAB}{\text{area } OABD} = \frac{h\left[(0h)^k + (1h)^k + (2h)^k + \cdots + (nh)^k\right]}{h\left[(nh)^k + (nh)^k + (nh)^k + \cdots + (nh)^k\right]}$$

$$= \frac{0^k + 1^k + 2^k + \cdots + n^k}{n^k + n^k + n^k + \cdots + n^k} . \tag{5.4}$$

Experimenting with (5.4) for values of k up to 10, Wallis discovered what he believed were unambiguous patterns. For example, when $k = 1$ we have

$$n = 1 : \quad \frac{0+1}{1+1} = \frac{1}{2} , \quad n = 2 : \quad \frac{0+1+2}{2+2+2} = \frac{1}{2} , \quad \text{etc.}$$

Trying $k = 2$ we see

$$n = 1: \quad \frac{0^2 + 1^2}{1^2 + 1^2} = \frac{1}{2} = \frac{1}{3} + \frac{1}{6 \cdot 1} \ ,$$

$$n = 2: \quad \frac{0^2 + 1^2 + 2^2}{2^2 + 2^2 + 2^2} = \frac{5}{12} = \frac{1}{3} + \frac{1}{6 \cdot 2} \ ,$$

$$n = 3: \quad \frac{0^2 + 1^2 + 2^2 + 3^2}{3^2 + 3^2 + 3^2 + 3^2} = \frac{14}{36} = \frac{1}{3} + \frac{1}{6 \cdot 3} \ , \text{etc.}$$

From observations like these, Wallis concluded that

$$\frac{0^2 + 1^2 + 2^2 + \cdots + n^2}{n^2 + n^2 + n^2 + \cdots + n^2} = \frac{1}{3} + \frac{1}{6n} \ . \tag{5.5}$$

As n increases, the term $1/6n$ vanishes. Thus, the ratio in (5.4) for the parabola $y = x^2$ is $1/3$. Because the area of $OABD$ is a^3, the area of region OAB is $(1/3)a^3$. (This matches Cavalieri's conclusion in section 4.1.)

Wallis went on, showing that when $k = 3$,

$$\frac{0^3 + 1^3 + 2^3 + \cdots + n^3}{n^3 + n^3 + n^3 + \cdots + n^3} = \frac{1}{4} + \frac{1}{4n} \ , \tag{5.6}$$

and again the last fraction vanishes "when n becomes infinite," as Wallis put it. Collecting these results, Wallis ultimately concluded that for any integer $k \geq 1$,

$$\frac{0^k + 1^k + 2^k + \cdots + n^k}{n^k + n^k + n^k + \cdots + n^k} = \frac{1}{k+1} \quad \text{when } n \text{ becomes infinite.} \tag{5.7}$$

From this and the fact that the area of $OABD$ is a^{k+1}, Wallis had discovered that

$$\begin{array}{c} \text{area under } y = x^k \\ \text{between } x = 0 \text{ and } x = a \end{array} = \frac{a^{k+1}}{k+1} \ . \tag{5.8}$$

This result suffices for Mercator's quadrature of the hyperbola, although Wallis was by no means finished. He then relaxed the stipulation that k be an integer, and exercise 5.1 tells this story.

Mercator exploited Wallis's result (5.8) by shifting the hyperbola $y = 1/x$ one unit to the left, as in Figure 5.7. He divided the horizontal OA into n equal parts of length h so that $a = nh$. Constructing rectangles on these subdivisions exhausts most of the region $OABC$ under the hyperbola so that when n becomes infinite,

$$\text{area of region } OABC = h \left(\sum_{k=1}^{n} \frac{1}{1 + kh} \right). \tag{5.9}$$

Each of the fractions in (5.9) expands by means of (2.7), and adding the expansions column by column gives us

$$\text{area of region } OABC = hn - h\big[h + 2h + 3h + \cdots + nh\big]$$

$$+ h\big[h^2 + (2h)^2 + (3h)^2 + \cdots + (nh)^2\big]$$

$$- h\big[h^3 + (2h)^3 + (3h)^3 + \cdots + (nh)^3\big]$$

$$+ \cdots .$$

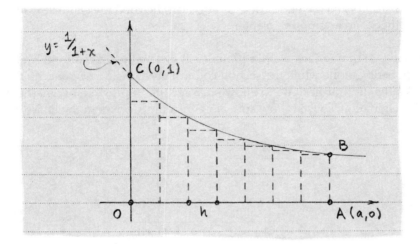

Figure 5.7. A hyperbola shifted one unit to the left opens the door for Wallis's method.

Comparing this expression to (5.4), we see that all terms but the first are the areas under curves $y = x^k$ as explored by Wallis in his pursuit of (5.8). Further, de Sarasa's observation (5.1) tells us that the area of region $OABC$ is a logarithm. The last piece of the puzzle locks into place, not only revealing that

$$\log(1 + a) = a - \frac{a^2}{2} + \frac{a^3}{3} - \frac{a^4}{4} + \cdots \tag{5.10}$$

but also giving the solution[2] to the alternating signed series (5.3):

$$\log 2 = 1 - \frac{1}{2} + \frac{1}{3} - \frac{1}{4} + \cdots . \tag{5.11}$$

This is a fittingly dramatic conclusion to an argument that establishes connections among lines of thought that spanned the previous twenty centuries. Other thinkers, inspired by these discoveries, were poised to reveal further connections, new results that would dethrone those in this chapter entirely.

5.5 Furthermore

5.1 **Wallis assigns a meaning to fractional exponents.** Most mathematical notation undergoes a lengthy evolution in which scholars both simplify and clarify it until they like it. Exponents emerged from just such a process. John Wallis was one of the first to denote and interpret exponents in the way we do today.

Wallis established his formula (5.8), which gives the area under $y = x^k$ from $x = 0$ to $x = a$, by experimenting with integer values of k. He now possessed a

---

[2]Exercise 6.6(a) explains why the base of this logarithm is the constant e.

key that unlocked a well-known door; boldly, he tried the same key on a door few others had attempted to enter.

(a) Table 3 summarizes and simplifies (5.7), where the first column lists the exponent k alone while the second column gives the corresponding fraction. Wallis did not hesitate to allow such results to provoke his mathematical imagination, so we follow suit. Note the empty boxes in the table; if we fill

$k = $ exponent	$\dfrac{1}{k+1}$
0	$1/(0+1)$
$\square$	$\square$
1	$1/(1+1)$
2	$1/(2+1)$
3	$1/(3+1)$
4	$1/(4+1)$
$\vdots$	$\vdots$

Table 3. The first few values of k for Wallis's result (5.7).

in the exponent box with $1/2$, then what seems likely to belong in the other box?

(b) Figure 5.8 shows the curve $y = x^2$ up to the point $(1, 1)$. Why do we know that the area of region OBC equals $2/3$?

(c) Why do these observations suggest that we interpret an expression like $m^{1/2}$ as $\sqrt{m}$?

(d) Can these same thoughts lead us to interpret $m^{1/3}$ as $\sqrt[3]{m}$?

5.2 Wallis verifies his ratio. Wallis validated his conclusion (5.5) by appealing to (2.1).

(a) Do as he did, and then verify (5.6) in a similar way.

(b) Try the case that 'follows' (5.6), that is, for the curve $y = x^4$.

(c) Wallis played with these ratios further, proving statements like

$$\frac{0^2 + 2^2 + 4^2 + \cdots + (2n)^2}{(2n)^2 + (2n)^2 + (2n)^2 + \cdots + (2n)^2} = \frac{1}{3}$$

when n becomes infinite. What simple algebra supports his conclusion? How far can you generalize this sort of observation?

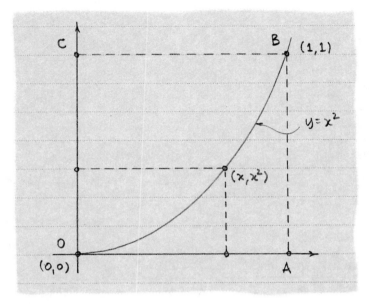

Figure 5.8. The area of region OBC helps to suggest a meaning for the exponent $1/2$.

5.3 **Wallis proposes a meaning for negative exponents.** The scholar **William Oughtred** (England, born 1574) proposed something like negative exponents in his work on logarithms, but Wallis was the first to study the matter. Table 3 inspired him. Each row represents a particular case of (5.7), and Wallis considered what happens when we 'divide' one row by another.

For example, we could divide each term in

$$0^3 + 1^3 + 2^3 + \cdots + n^3, \text{ where } k = 3$$

by each respective term in

$$0^2 + 1^2 + 2^2 + \cdots + n^2, \text{ where } k = 2$$

to get

$$0^1 + 1^1 + 2^1 + \cdots + n^1, \text{ where } k = 1.$$

Dividing in this way corresponds to subtracting the values of k.

(a) With this in mind, what do you think that Wallis concluded in the case where $k = -1$?

(b) Figure 5.9 connects the idea that $x^{-1} = 1/x$ to the geometry that underlies Figure 5.7. Wallis concluded that the area under the curve $y = x^{-1}$ between $x = 0$ and $x = 1$ is infinite. The harmonic series (3.12) plays the key role. Reproduce his argument.
Historical note. The row of Table 3 that corresponds to the curve $y = x^{-1}$ would read -1 and $1/(-1 + 1)$, or -1 and $1/0$. Wallis took from this that

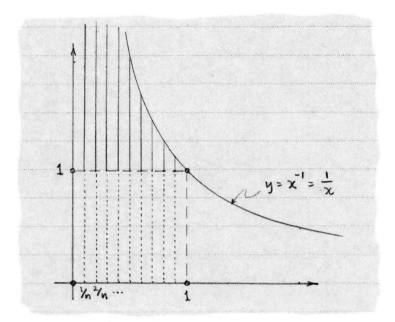

Figure 5.9. Interpreting x^{-1} as $1/x$ led Wallis to consider the area beneath a hyperbola between $x = 0$ and $x = 1$.

1/0 equals ∞, a symbol he invented to mean 'infinity'. He then argued that a row reading -2 and $1/(-2 + 1)$ corresponds to the curve $y = x^{-2}$, which encloses even more area than does $y = x^{-1}$ between $x = 0$ and $x = 1$. From this he speculated that

$$\frac{1}{-1} > \infty.$$

5.4 **Briggs calculates the (common) logarithm of 2.** One field of human endeavor often prompts advances in another. In the late 1500s, for example, better ships allowed for riskier travel, and captains who ventured across the Atlantic or around the extremities of the continents cared deeply about precise navigation. To pinpoint the locations of their ships at sea, navigators relied on trigonometry. Their calculations required the multiplication and division of coordinates having many decimal places.

As we discussed in section 5.2, logarithms offer a way to turn multiplication problems into additions (and divisions into subtractions). A logarithmic table like that in Table 1, however, cannot provide much precision. **John Napier** (Scotland, born 1550) spent a good deal of his life calculating tables of logarithms intended to provide accuracy to seven decimal places. (As decimal places were not a commonly accepted notation during Napier's life, his work on logarithms helped to convince scholars to adopt it.)

Just before his death, Napier hosted English mathematician **Henry Briggs** (born 1561), whose interest in Napier's work led him to create a table of logarithms of his own. This question outlines his approach.

(a) Consider Table 4, a variation on Table 1 where the arithmetic sequence has been replaced by one that begins with 3 rather than 0. We wish to translate

x	$\log x$
1	3
2	4
4	5
8	6
16	7
32	8
64	9

Table 4. A slightly altered version of Table 1.

the multiplication of two numbers a and b in the x column into the addition of two numbers in the $\log x$ column as we did before, using

$$\log ab = \log a + \log b . \qquad (5.12)$$

For example, if $a = 4$ and $b = 8$, then we observe that

$$\log 4 + \log 8 = 5 + 6 = 11 ,$$

and we use Table 4 to find the value of x such that $\log x = 11$. Unfortunately, the table (if extended) gives $x = 256$, which is not the product of 4 and 8.

The trouble stems from our use of an arithmetic sequence that begins with 3 and thus requires a correction; our patched-up version of (5.12) is

$$\log ab = \log a + \log b - \log 1 ,$$

which now works. Briggs elected to assign $\log 1 = 0$ in part to make the simpler version (5.12) valid. He also set $\log 10 = 1$, a choice that gives a special status to the number 10 in these calculations; we say that 10 is the *base* for the logarithms of Briggs. Ordinarily, we write the base as a subscript, as in $\log_{10} 10 = 1$, but by omitting the base, we imply that we are using the *common logarithm* with base 10.

Briggs added to his short list of logarithms in Table 5 by calculating the square root of 10. Why does (5.12) justify his conclusion that

$$\log(10^{1/2}) = 1/2 ?$$

x	$\log x$
10	1
$10^{1/2} \approx 3.16228$	$1/2$
1	0

Table 5. The first three steps in Briggs's approach.

(b) The square root of the square root of 10 is

$$(10^{1/2})^{1/2} = 10^{1/4} \approx 1.77828,$$

a number Briggs calculated to six times as many decimal places using a method of his own devising. (Incidentally, he erred in the nineteenth decimal place, thanks surely to the difficulty of finding such precise square roots entirely 'by hand' on paper.) Continuing this process, Briggs created a list much like the one in Table 6. Each entry in the $\log x$ column is half of

x	$\log x$
10	1
$10^{1/2} \approx 3.16228$	$1/2 = 0.5$
$10^{1/2^2} \approx 1.77828$	$1/2^2 = 0.25$
$10^{1/2^3} \approx 1.33352$	$1/2^3 = 0.125$
$10^{1/2^4} \approx 1.15478$	$1/2^4 = 0.0625$
$10^{1/2^5} \approx 1.07461$	$1/2^5 = 0.03125$
$10^{1/2^6} \approx 1.03663$	$1/2^6 \approx 0.01563$
$10^{1/2^7} \approx 1.01815$	$1/2^7 \approx 0.00781$
$10^{1/2^8} \approx 1.00904$	$1/2^8 \approx 0.00391$
$10^{1/2^9} \approx 1.00451$	$1/2^9 \approx 0.00195$
$10^{1/2^{10}} \approx 1.00225$	$1/2^{10} \approx 0.00098$
$10^{1/2^{11}} \approx 1.00112$	$1/2^{11} \approx 0.00049$
1	0

Table 6. The table of logarithms grows more complete.

the entry above it, of course, while the entries in the x column do not behave this way. However, if we subtract 1 from the entries in the x column, the ratio of each to the one before comes quite close to $1/2$ at the bottom; for example,

$$\frac{10^{1/2^{11}} - 1}{10^{1/2^{10}} - 1} \approx \frac{0.00112}{0.00225} \approx 0.49778 \approx \frac{\log 10^{1/2^{11}}}{\log 10^{1/2^{10}}} . \tag{5.13}$$

In short, the decimal parts of entries in the x column become roughly proportional to the corresponding logarithms in the other column.

Figure 5.10 shows the situation geometrically. On the vertical (at points A, B, C) are the decimal parts of the three entries near the bottom of the 'x'

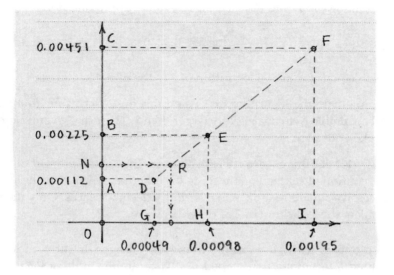

Figure 5.10. Because OA is roughly half of OB and OB roughly half of OC, segments DE and EF nearly create a straight line.

column in Table 6. On the horizontal (at points G, H, I) are the corresponding logarithms. The length of OG is *exactly* half that of OH, while the length of OA is *roughly* half of OB. Similarly, we see that $OH/OI \approx OB/OC$. Thus, Briggs reasoned that the decimal part of any number N that falls between 1.00112 and 1.00225 would *roughly* correspond to a point R on the segment DE. The algebraic version of this reasoning imitates (5.13) like so:

$$\frac{10^{1/2^{11}} - 1}{N - 1} \approx \frac{\log 10^{1/2^{11}}}{\log N} \implies \log N \approx \frac{(N-1)(0.00049)}{0.00112} . \tag{5.14}$$

We may choose

$$N = 1.024^{1/2^4} = \left(\frac{2^{10}}{1000} \right)^{1/2^4} \approx 1.00148$$

and capture $\log 2$ via these calculations:

$$\begin{aligned}
\log N &= \log \left(\frac{2^{10}}{1000} \right)^{1/2^4} \\
&= \frac{1}{2^4} \left(\log(2^{10} \cdot 10^{-3}) \right) \\
&= \frac{1}{16} \left(10 \log 2 + (-3) \log 10 \right) .
\end{aligned}$$

Solving for $\log 2$, we have

$$\log 2 = \frac{16 \log N + 3}{10} . \tag{5.15}$$

Combining (5.14) and (5.15), we arrive at the approximation

$$\log 2 \approx 0.301028,$$

which equals the actual logarithm of 2 to the fifth decimal place.

Approximate the logarithm of 3 using this approach. Your first task is to find a suitable N that is slightly larger than 1. There are several candidates, so take your pick.

Historical note. Briggs was not satisfied with the "rough proportion" in (5.13), and continued taking square roots until he had calculated the *fifty-fourth* successive square root of 10 to over *thirty* decimal places. Only then did he calculate the logarithm of 2 as described here.

5.5 **Briggs pioneers work on interpolation.** Henry Briggs, introduced in exercise 5.4, created a technique that helped him generate the table of logarithms described in that same exercise. (It is not necessary to have worked through that exercise to benefit from this question.)

(a) Painstaking work allowed Briggs to calculate $\log 2$ and $\log 3$ (as outlined in exercise 5.4), and one use of (5.12) gave him

$$\log 4 = \log(2 \cdot 2) = \log 2 + \log 2 \approx 0.6020599913$$

as well. Between these three logarithms lie others that Briggs approximated by *interpolation*. The prefix *inter-* means 'between', and the ending *-polate* uses the same root as *polish*, as in 'to shine or improve'. To 'interpolate' between two known values, therefore, is to produce an approximation that becomes better the longer we work at it.

Figure 5.11 both motivates and illustrates the method of interpolation. We see that the points A, B, C, which correspond to the values $\log 2$, $\log 3$, $\log 4$, respectively, do not lie in a straight line. If we pay attention to this, then we might improve our approximation should we attempt to interpolate a value for, say, $\log 2.5$. Rather than naively guess that $\log 2.5$ corresponds to a point on AB, we can involve point C by imagining a curve (the dotted curve in the figure) that passes through all three points A, B, C. It seems reasonable to assign $\log 2.5$ to a point on this curve, because the curve depends on three known logarithms rather than just two.

To interpolate $\log 2.5$, Briggs created Table 7 to organize his calculations. He found the 'first differences' of the known logarithms; these appear in Figure 5.11 as the slopes of AB and BC. He called the difference of these differences the 'second difference', noting that the more this number deviates from zero, the more central is its role in the interpolation.

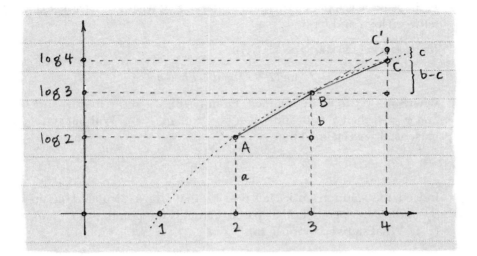

Figure 5.11. This figure motivates Briggs's method of interpolation.

		first differences		second difference
$\log 4 \approx 0.6020599913$	$\rangle$			
		0.1249387366		
$\log 3 \approx 0.4771212547$	$\rangle$		$\rangle$	-0.0511625225
		0.1760912591		
$\log 2 \approx 0.3010299957$				

Table 7. A little bookkeeping for interpolating $\log 2.5$.

In Figure 5.11, point C' lies on the intersection of the vertical line $x = 4$ and the extension of AB. With this in mind, the 'differences' in Table 7 appear in the figure as

$$a = \log 2 ,$$
$$b = \text{the first difference } \log 3 - \log 2 ,$$
$$b + c = \text{the first difference } \log 4 - \log 3 ,$$
$$c = \text{the second difference} .$$

Note that c is negative thanks to the bend of the dotted curve.

Let k be the positive number so that $\log(2+k)$ is our target. (In our example, we wish to find $\log 2.5$, so $k = 0.5$.) The values $k = 0, 1, 2$ place us at the three logarithms we know:

$$k = 0 : \quad \log 2 = a ,$$
$$k = 1 : \quad \log 3 = a + b ,$$
$$k = 2 : \quad \log 4 = a + 2b + c . \tag{5.16}$$

Perhaps there is a pattern here to exploit so that we may interpolate $\log 2.5$ between $\log 2$ and $\log 3$.

It appears safe to start guessing with

$$\log(2 + k) = a + kb + qc \,,$$

where q is yet to be determined. From (5.16), we know that when $k = 0$ and $k = 1$, then $q = 0$, and when $k = 2$, then $q = 1$. Without stating his reasons, Briggs claimed that

$$q = \frac{1}{2}k(k-1) \,.$$

Indeed, this solution works. It is possible that Briggs calculated this answer by finding the parabola that passes through the points $(0,0)$, $(1,0)$, and $(2,1)$. What steps accomplish this?

(b) Thus, the interpolation formula is

$$\log(2 + k) \approx a + kb + \frac{1}{2}k(k-1)c \,, \tag{5.17}$$

where we find a, b, c using differences like those in Table 7. Use (5.17) to approximate $\log 2.5$. Try another value, such as $\log 2.1$, as well.

(c) If we wished to (more roughly) approximate $\log 2.5$ by locating the corresponding point on AB in Figure 5.11, how would we modify (5.17)?

5.6 **James Gregory extends interpolation.** In his method of interpolation, as given by (5.17), Briggs used three points of data, at equally-spaced intervals, to make predictions for other values within those intervals. Whenever the second difference was not small, however, Briggs extended his method to use more points of data.

James Gregory (Scotland, born 1638) explored this approach as well, using more than three points of data to make what he assumed were even better predictions. This exercise explores one step in this process: specifically, if we have four points of data, then how should we alter (5.17) to make use of this?

Figure 5.12 revisits Figure 5.11 in a more general setting. The three known data points are labeled A, B, C, with vertical heights $a, a+b, a+2b+c$ corresponding to values $x, x + h, x + 2h$ along the horizontal. Briggs's formula (5.17) handles this situation; the interpolated height of any value $x + kh$ (where k is a number between 0 and 2, to keep the value in the appropriate interval) is

$$a + kb + \frac{1}{2}k(k-1)c \,. \tag{5.18}$$

By introducing a fourth data point D that corresponds to the value $x + 3h$, we hope to improve on the prediction given by (5.18). In exercise 5.5, we saw that

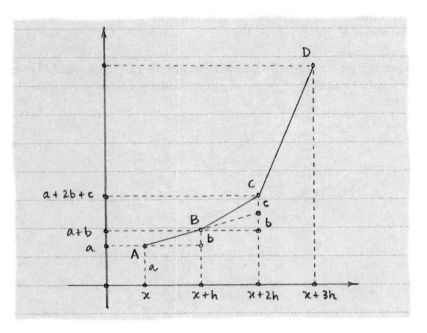

Figure 5.12. This generalized version of Figure 5.11 indicates how James Gregory extended the method of interpolation.

each data point corresponded to one term in (5.17), so we suspect that a fourth point will allow us to add a fourth term to the end of (5.18).

(a) Table 8 does for Figure 5.12 as Briggs did in Table 7. Fill in the empty boxes; the rest of the entries in the table reside in Figure 5.12. Explain your reasoning.

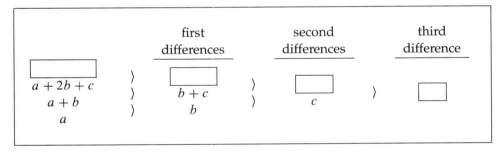

Table 8. We must fill in the four empty boxes.

(b) Recreate Figure 5.12 and extend the dashed lines that pass through A and B, and extend BC, until all of these lines intersect the vertical line at D, breaking this vertical line into five segments. If we say that the length of the topmost segment is $c + d$, then what are the lengths of the other four segments in terms of a, b, c? Use your results from 5.6(a) to help.

(c) With these thoughts in mind, determine what term ought to be added to the end of (5.18).

(d) Simply by pattern-finding, guess the fifth term, which would be useful if we knew a fifth data point. If you introduce any new variables, explain them.

6

The Fundamental Theorem of Calculus

Evidence mounted in the late 1600s that efforts to understand quadrature and attempts to quantify instantaneous velocity could be unified in a single theory. This link was confirmed by Isaac Newton of England and Gottfried Leibniz of Germany. For this achievement, we honor them as the discoverers of calculus.

6.1 Newton links quadrature to rate of change

We may view an object's velocity as the *rate of change* of its distance. Not all curves describe distance, but many curves allow for tangent lines. Thus, we speak of the rate of change of a curve from this point forward, unless the situation specifically describes motion.

We finally meet **Isaac Newton** (England, born 1642), a thinker whose broad interests could have allowed any of several earlier introductions. He approximated π, identified undiscovered series, pioneered work in interpolation, and determined the mathematics that underlies gravity. In exercise (6.5) we will see how he overcame the difficulty of expanding expressions like $(a + e)^{1/2}$ using infinite sums. For that discovery alone, Newton would have earned a place in this story. His insights about quadrature and rate of change, however, promote him to central character, as reflected in the title given to his discovery: *the fundamental theorem of calculus.*

Here is one side of this story: in Figure 6.1, curve ABC increases as we move left to right. Corresponding to motion along this curve is motion along the horizontal from A to D to E and so on. Relate each point on the horizontal to the point on the curve above it, as depicted by segments DB and EF. As the segments sweep to the right, the area they enclose grows larger. Region ADB is the area enclosed by DB, and if distance DE is very small, then we may think of region $DEFB$ as a small increase in that area.

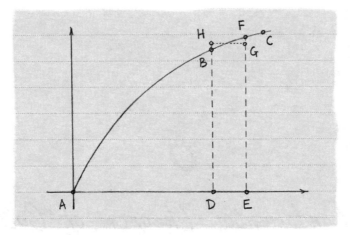

Figure 6.1. Newton thought of the quadrature of ABC as if it were a changing quantity.

Newton's insight here was to treat the area beneath curve ABC as a changing quantity. Think of ABC as representing the velocity of an object; as previously mentioned (in section 3.1), the quadrature of such a curve gives the distance traveled by the object. Region $DEFB$ therefore represents a tiny change in distance over a tiny change DE in time. If ABC changes rapidly at time D, then the area beneath it also changes rapidly. So we might suspect, as Newton did, that a curve is linked to the rate of change of the area beneath it.

Newton argued for this link by asserting that given any such curve ABC, and any tiny change DE along the horizontal, then we can find rectangle $DEGH$ having area equal to that of $DEFB$, as in Figure 6.1. Thus the tiny change in the area of region ADB equals the area of $DEGH$, which equals $DH \cdot DE$. Because DE is a tiny change along the horizontal, we have

$$DH = \frac{\text{change in area } ADB}{\text{change in horizontal } DE} . \tag{6.1}$$

In the context where ABC represents the velocity of an object, we can translate (6.1) to

$$DH = \frac{\text{change in distance during time } DE}{\text{change in time } DE} . \tag{6.2}$$

Now hold point D fixed and slide point E closer and closer to it. As DE becomes "infinitely small", as Newton put it, so does DH become equal to DB. These observations allow us to translate (6.2) to

$$DB = \text{the rate of change in the area of } ABD \text{ at the point } D \ .$$

But DB is the height of curve ABC at D along the horizontal; thus, Newton has uncovered a link between a curve, its quadrature, and the rate of change of the quadrature:

> The height of a curve at a point along the horizontal equals
> the rate of change of the quadrature of the curve at that point.

(6.3)

This statement alone does not show the back-and-forth nature of quadrature and rate of change. Newton provided another argument that shows we may swap the concepts of quadrature and rate of change and arrive at a similar truth.

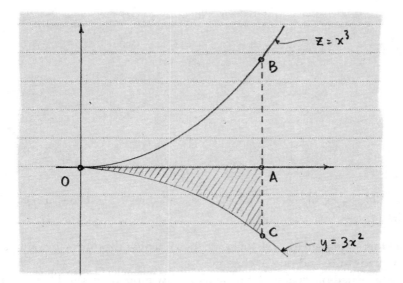

Figure 6.2. At any point, the height of y equals the slope of the tangent line at that point on z. (Note that y is reflected across the horizontal axis from its usual position, for the sake of clarity.)

6.2   *Newton reverses the link*

In this version, Newton used the curves $y = 3x^2$ and $z = x^3$, shown by others (e.g., Wallis in (5.8) with $k = 2$) to have a particular relationship: at any point, the height of y equals the slope of the tangent line at that point on z. Newton displayed the curves as shown in Figure 6.2, with z oriented as modern readers are accustomed, and y reflected down across the horizontal, which seems strange to modern eyes.

Focusing upon a point A on the horizontal, we locate points B and C on the respective curves. Newton desired to show that the quadrature of y as denoted by the shaded region OAC equals the height AB of the curve z. Success would allow him to conclude that

> The height of a curve at a point along the horizontal equals the
> quadrature of the rate of change of the curve up to that point.

(6.4)

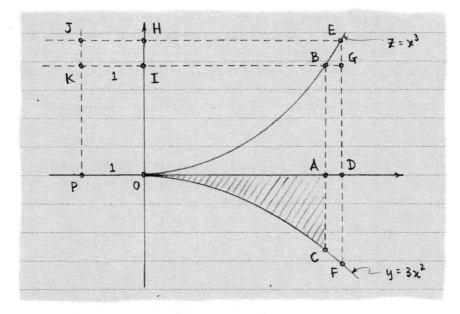

Figure 6.3. Newton's goal was to show that the shaded region OAC has area equal to the length AB.

Compare this statement to (6.3); together, they uncover the mirror-like relationship between quadrature and rate of change.

Newton reached his goal with the help of a few more points and lines, a familiar argument, and the usual assumptions about the infinite. In Figure 6.3, point D is placed so that AD is infinitely small. Points E and F lie on the curves above and below D, and G sits on DE so that BG and DE are perpendicular. Point P is introduced on the horizontal so that OP has length 1, and upon OP we construct a rectangle $OPJH$ with PJ parallel to OH and JHE parallel to both KIB and POA.

The rest of the argument is an analysis of this figure. The slope of the tangent line to z at point B is equal to not only EG/BG, thanks to AD being infinitely small, but also AC, thanks to y giving the slope of z at any point. Thus,

$$AC = EG/BG \implies EG = AC \cdot BG$$
$$\implies HI = AC \cdot AD$$
$$\implies \text{area of rectangle } KIHJ \approx \text{area of region } ADFC \ .$$

Newton treated this approximation as an equality because AD is infinitely small. The equality of these matching areas holds for any point between O and A, so

$$\text{area of rectangle } POIK = \text{area of region } OAC \ .$$

Because OP has length 1, the area of $POIK$ is the height AB of the curve z. Thus, the height of the curve z at the particular point A equals the quadrature (up to

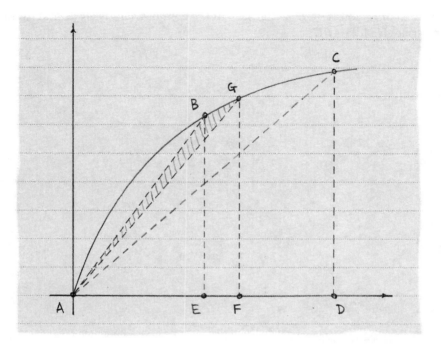

Figure 6.4. Leibniz subdivided region ADC into $\triangle ADC$ and "infinitely many" triangles like ABG.

that point) of the curve y that represents the rate of change of z. This is exactly the statement in (6.4) that Newton wished to prove.

6.3 *Leibniz discovers the transmutation theorem*

This 'inverse' relationship between quadrature and rate of change was confirmed independently by **Gottfried Leibniz** (Germany, born 1646). As Newton did, Leibniz found a geometric figure that involved "infinitely small" quantities. Exercise 7.3 explores his work on the fundamental theorem; here, we see how Leibniz used a similar argument to transform certain difficult quadrature problems into simpler ones.

Curve ABC in Figure 6.4 bounds region $ABCD$ and Leibniz wished to find its quadrature. He cut the region into two pieces with segment AC, so

$$\text{area of region } ABCD = \text{area of } \triangle ACD + \text{area of region } ACB . \qquad (6.5)$$

Finding the area of a triangle is simple, so Leibniz immediately focused on the area of region ACB, exhausting the area with infinitely thin triangles as follows. Point B on the curve is related to point E on the horizontal, and point F (matched with point G on the curve) is an infinitely small distance from E. We create (shaded) triangle ABG (which is both infinitely small and not quite a triangle). Each point on curve ABC between A and C is associated with such a triangle, so we can think of the area of region $ABCD$ as the sum of the area of triangle ACD and infinitely many triangles like ABG.

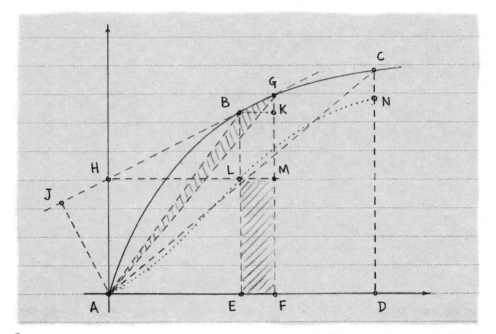

Figure 6.5. Leibniz showed that the area of $\triangle ABG$ equals half the area of rectangle $EFML$.

Leibniz discovered a way to use the rate of change of curve ABC, as given by the slope of the tangent line to the curve, to find the area of these infinitely small triangles. Because EF is infinitely small, segment BG acts like a tangent line to ABC at point B. Extend BG as in Figure 6.5 so that it crosses the vertical above A at H. Let AJ be perpendicular to this line, and note that AJ is the height of triangle ABG (with base BG). Thus,

$$\text{area of } \triangle ABG = \frac{1}{2} \, AJ \cdot BG \, . \tag{6.6}$$

Locating K so that BK is perpendicular to FG, we create the infinitely small $\triangle BKG$. Because AH is parallel to EB, we know $\angle JHA = \angle HBE$. Now angles EBK and AJH are both right angles, so $\angle KBG = \angle JAH$. Thus, triangles BKG and AJH are similar, so

$$\frac{BK}{BG} = \frac{AJ}{AH} \Longrightarrow AJ \cdot BG = BK \cdot AH \, . \tag{6.7}$$

Leibniz used the term *characteristic triangle* to describe an infinitely small triangle like BKG. The word *characteristic* describes a feature of some object that is distinctive; in the case of a curve, its steepness at each point helps to distinguish it from other curves. The hypotenuse of a characteristic triangle depicts the curve's steepness.

To continue with the argument: draw HLM parallel to AEF. Combining (6.6) with (6.7) gives

$$\text{area of } \triangle ABG = \frac{1}{2} BK \cdot AH = \frac{1}{2} \text{ area of } EFML .\qquad(6.8)$$

Leibniz has transformed a triangular area to half of a rectangular area with the help of a tangent line; the link between quadrature and rate of change begins to suggest itself.

Now imagine point E (and its companion F) in motion as E travels from A to D. For each place E occupies, there is a corresponding point L (which is defined by H, which is defined by tangent line BG). Plotting all such points L results in the dotted curve ALN in Figure 6.5. The corresponding (infinitely thin) rectangles $EFML$ will sweep through the region $ALND$; the error just above $EFML$ is insignificant thanks to its infinite thinness. Now as E moves from A to D, the corresponding point B moves from A to C, so triangles ABG sweep through the region $ALCB$. Thus, by (6.8), we know that

$$\text{area of region } ACB = \frac{1}{2} \text{ area of region } ALND .\qquad(6.9)$$

Finally, equations (6.5) and (6.9) together yield

$$\text{area of region } ABCD = \text{area of } \triangle ACD + \frac{1}{2} \text{ area of region } ALND .\qquad(6.10)$$

Leibniz called this equation his "transmutation" theorem. At first glance, his transmutation of the area of region $ABCD$ has not accomplished much: although the area of triangle ACD is easy to find, the area of region $ALND$ may be no simpler to calculate than the area of the original region $ABCD$. Luckily, it often *is* simpler, and the next section offers an example.

6.4   Leibniz attains Jyesthadeva's series for π

Leibniz found many geometric problems that succumbed to his transmutation theorem (6.10). One such problem ties together much of what we have been studying, and relates to the quadrature of a quarter-circle. Leibniz used his theorem to find a new proof of Jyesthadeva's formula (2.13):

$$\frac{\pi}{4} = 1 - \frac{1}{3} + \frac{1}{5} - \frac{1}{7} + \frac{1}{9} - \cdots .$$

Figure 6.6 shows a quarter-circle $ABCD$ of radius 1 with a typical point E at distance x from A along the horizontal. Line BH is tangent to the circle at point B. All points are labeled to correspond to their counterparts in Figure 6.5. Curve ALN plays the same role, too, only here N coincides with C. For convenience, let $EB = y$ and $EL = z$.

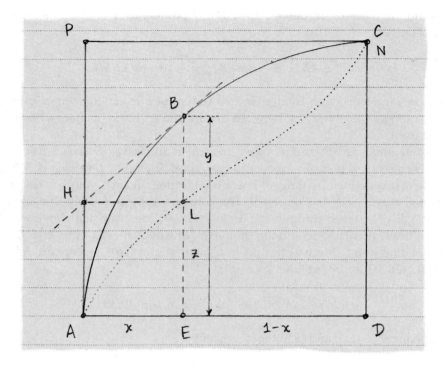

Figure 6.6. Leibniz found a quadrature of the quarter-circle using his transmutation theorem.

The transmutation theorem (6.10) gives us

$$\text{area of quarter-circle } ABCD = \text{area of } \triangle ADC + \frac{1}{2} \text{ area of region } ALND. \quad (6.11)$$

Finding the area of region $ALND$ is not simple, and the solution by Leibniz was ingenious. First, observe that

$$\text{the slope of } BH = \frac{BL}{HL} = \frac{y - z}{x} \implies z = y - x \cdot (\text{the slope of } BH). \quad (6.12)$$

We establish a relationship between x and z with the help of (6.12) if we note that $y = \sqrt{2x - x^2}$ (draw BD and apply the Pythagorean Theorem to triangle BDE) and thus the slope of BH is $(1 - x)/\sqrt{2x - x^2}$ (because BD is perpendicular to BH). Hence,

$$z = \sqrt{2x - x^2} - \frac{x(1 - x)}{\sqrt{2x - x^2}} \implies x = \frac{2z^2}{1 + z^2}.$$

With this relationship established, we can make headway with (6.11) as follows:

$$\text{area of quarter-circle } ABCD = \text{area of } \triangle ADC + \frac{1}{2} \text{ area of region } ALND$$

$$= \frac{1}{2} + \frac{1}{2}(1 - \text{area of region } ALNP)$$

$$= 1 - \frac{1}{2}(\text{area of region } ALNP). \quad (6.13)$$

Leibniz considered the area of region $ALNP$ as the sum of all of the line segments like HL contained therein. Now HL has length

$$x = \frac{2z^2}{1 + z^2}$$

$$= 2z^2 \frac{1}{1 + z^2}$$

$$= 2z^2(1 - z^2 + z^4 - z^6 + z^8 - \cdots)$$

$$= 2(z^2 - z^4 + z^6 - z^8 + z^{10} - \cdots)$$

where the next-to-last equality is justified by (2.7). So to calculate the area of region $ALNP$, Leibniz summed

$$2(z^2 - z^4 + z^6 - z^8 + z^{10} - \cdots)$$

as z increases from 0 to 1. Each term in the series contributes to the total in turn; starting with z^2, we sum from $z = 0$ to $z = 1$ as Wallis demonstrated in (5.8), reproduced (with amendments) here:

$$\text{area under parabola } y = z^k \text{ between } z = 0 \text{ and } z = 1 = \frac{1^{k+1}}{k + 1}.$$

Thus, the term z^2 adds $1/3$ to the total, while the term z^4 subtracts $1/5$, and so on. This allowed Leibniz to reach his grand conclusion:

$$\text{area of quarter-circle } ABCD = 1 - \frac{1}{2}(\text{area of region } ALNP)$$

$$= 1 - \frac{1}{2}\left(2\left(\frac{1}{3} - \frac{1}{5} + \frac{1}{7} - \frac{1}{9} + \cdots\right)\right)$$

$$= 1 - \frac{1}{3} + \frac{1}{5} - \frac{1}{7} + \frac{1}{9} - \cdots .$$

As the area of the quarter-circle $ABCD$ is $\pi/4$, we see that Leibniz has rediscovered Jyesthadeva's wonderful formula (2.13). Newton wrote of this discovery that it "sufficiently revealed the genius of its author, even if he had written nothing else."

Leibniz was by no means finished. While making his discoveries, he invented notation that turned complicated geometric arguments like those in this chapter into more straightforward manipulation of symbols. These efforts are an important part of our story as we continue.

6.5 *Furthermore*

6.1 **Pascal inspires Leibniz with his 'sum of sines.'** Leibniz credited **Blaise Pascal** (France, born 1623) with setting him on a path that led to the fundamental theorem. Within one of Pascal's investigations of circles (the one detailed in this exercise) was a geometric approach that Leibniz generalized to other curves.

(a) Pascal discovered a link between tangent lines and area that we will follow using the quarter-circle OAB in Figure 6.7. Draw any radius OD and a short segment LR that is tangent to the arc and has D as its midpoint. Segments LX, DC, RY are all perpendicular to OB, and K sits on LX so that LKR is a right angle. Pascal claimed that because a line tangent to a circle is perpendicular to the radius at the point of tangency, then $\triangle OCD$ is similar to $\triangle LKR$. Why is he correct?

(b) Pascal meant for us to view LR as equivalent to the arc of the circle that it shadows, as though the entire arc AB were composed of segments like LR. He granted that his claim is true only when the number of such segments "is infinite", but was careful to state that this is a shorthand way of saying something more mathematical: that the difference between the length of AB and the sum of the lengths like LR can be made as small as possible by increasing the number of segments like LR.

Using the similar triangles mentioned in 6.1(a), explain why we may conclude that

$$DC \cdot LR = XY \cdot OD .\qquad\qquad (6.14)$$

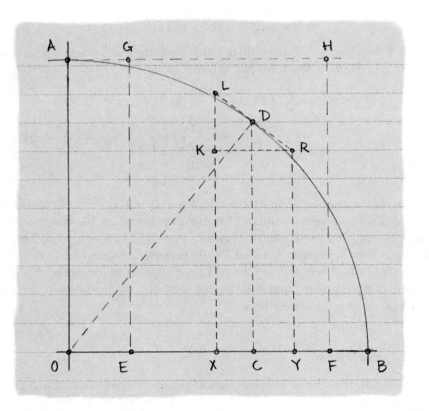

Figure 6.7. Pascal's triangle LKR prompted Leibniz to consider the 'characteristic triangle' mentioned in section 6.3.

Historical note. Equation (6.14) reflects the core idea of the fundamental theorem: the left-hand side, which involves a tangent to a curve, equals the right-hand side, an area that depends on the curve, specifically the radius of the circle. Pascal did not, however, make note of this link between tangency and quadrature.

(c) Pascal translated (6.14) much like this: Choose any point on the arc; its vertical height, multiplied by the length of its tangent segment, equals the length of the corresponding horizontal segment multiplied by the radius. Now Pascal summed both sides of (6.14), as if he were considering all of the vertical segments DC as they sweep from E to F in Figure 6.7. What reasoning supports his claim that this sum, for the right-hand side of (6.14), equals the area of rectangle $EFHG$?

(Note: The sum of the left-hand side of (6.14) is not so simple, for the height of DC changes as it sweeps through the portion of the quarter-circle above EF. We will return to this issue in exercise 8.4, when we have better notation.)

6.2 **Barrow anticipates the fundamental theorem.** Newton's mentor, **Isaac Barrow** (England, born 1630) drew a link between tangent lines and quadrature that foreshadowed Newton's fundamental theorem. A look at Barrow's argument should help solidify our study of that far-reaching result.

In the style of Figure 6.2, we place a curve y below the horizontal in Figure 6.8 and declare that it is 'increasing'; we use the mirror image of where we would normally depict y so that we can put a related curve z in that spot. At any point A, the height of AC is determined by the area of the region $OABP$. Comparing this to Newton's Figure 6.2, we note that Barrow *created* z to equal the area determined by y whereas Newton *proved* that z and y shared this relationship. So, despite the similarity of their drawings, Newton and Barrow had different aims.

(a) Barrow placed the point T on the horizontal so that the slope of TC equals the length of AB. So,

$$\frac{AC}{TA} = AB \implies AC = AB \cdot TA .$$

The importance of T becomes clearer as we proceed; for now, note that the area of region $OABP$ equals the area of the rectangle with sides AB and AT.

Barrow intended to prove that TC is tangent to z in the sense that C is the only point where TC and z intersect. His clever argument focused on each point J between T and C, showing that J always lies between point I on z and point K on AC (ensuring that TC does not intersect z to the left of C).

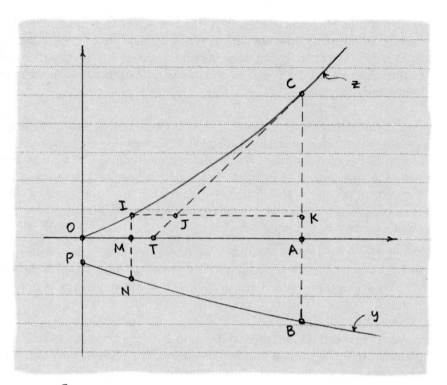

Figure 6.8. In $\triangle TAC$ lies the crux of the fundamental theorem.

Segment IK is parallel to the horizontal and IN is parallel to the vertical. With this in mind, why is it true that the height of KC equals the area of region $MABN$?

(b) Because y is increasing, we know that the area of region $MABN$ is less than the area of the rectangle with sides MA and AB. Why may we go on to conclude that $JK < IK$?

(c) This concludes Barrow's argument in the case where J is between T and C. What if J lies on the extension of TC through C? Draw a new figure, and write down the argument for this case.

6.3 **Maclaurin proves a special case of the mean value theorem.** This exercise primarily supports exercise 6.4 below, but also highlights a result called *the mean value theorem* that became more pertinent as scholars made calculus more rigorous. **Colin Maclaurin** (Scotland, born 1698) proved that

$$na^{n-1} < \frac{b^n - a^n}{b - a} < nb^{n-1} \tag{6.15}$$

when $a < b$ and n is an integer bigger than 1. The middle term of this inequality equals the slope of the secant line through the points on the curve $y = x^n$ where $x = a$ and $x = b$.

(a) Sketch the curve and secant line yourself. What line could you add to your sketch that would have slope na^{n-1}? How about nb^{n-1}? Does (6.15) make sense in light of your answers?

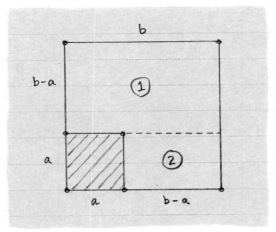

Figure 6.9. We may geometrically factor $b^2 - a^2$ by summing the areas of regions 1 and 2.

(b) No matter what n we choose, we may factor $b - a$ from $b^n - a^n$. For example, when $n = 2$ we have $b^2 - a^2 = (b-a)(b+a)$. Figure 6.9 provides a geometric basis for this conclusion; because $b^2 - a^2$ equals the sum of the areas of regions 1 and 2,

$$b^2 - a^2 = b(b-a) + a(b-a) = (b-a)(b+a) .$$

Draw a pair of cubes with sides a and b, oriented in a way similar to the squares in Figure 6.9, to help you argue that

$$b^3 - a^3 = (b-a)(b^2 + ab + a^2) .$$

(c) Based on these results for $n = 2$ and $n = 3$, conjecture how $b - a$ factors from $b^4 - a^4$. Check your idea algebraically.

(d) How do these observations lead you to the truth of (6.15)?

6.4 Maclaurin tackles the fundamental theorem algebraically. The arguments advanced by Newton and Leibniz for the fundamental theorem rely on geometry and "infinitely small" lengths like DE in Figure 6.1 and EF in Figure 6.4. Maclaurin sidestepped the "infinitely small" in his treatment of a portion of the fundamental theorem for a particular kind of curve. His approach provided hope that the tools of calculus might be freed of any reliance on metaphor.

Suppose a curve y has the property that the area under it up to any positive number x is equal to x^n. Maclaurin, aware of Wallis's result (5.8) and Newton's result (6.3), knew that this curve would *likely* be expressed as $y = nx^{n-1}$, but he wanted to *prove* this was so without resorting to phrases like "infinitely small".

(a) Figure 6.10 shows a sketch of y with a typical value x chosen along the horizontal. The value h equals some small positive quantity. Because the

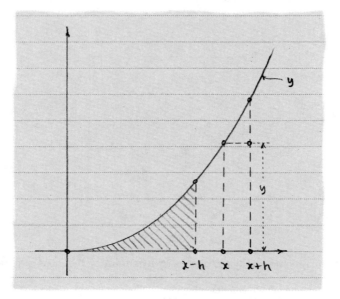

Figure 6.10. This figure justifies Maclaurin's claims about the areas mentioned in (6.16).

area under y up to any value x equals x^n, we know that (for example) the shaded area equals $(x - h)^n$. With this in mind, explain why

$$x^n - (x - h)^n < yh < (x + h)^n - x^n . \tag{6.16}$$

(b) Maclaurin substitutes $x - h$ for a and x for b in his result (6.15), concluding that

$$n(x - h)^{n-1}h < x^n - (x - h)^n .$$

Determine what we must substitute into (6.15) to show that

$$(x + h)^n - x^n < n(x + h)^{n-1}h .$$

(c) These results together demonstrate that

$$n(x - h)^{n-1} < y < n(x + h)^{n-1} . \tag{6.17}$$

Letting h vanish, we could satisify ourselves that $y = nx^{n-1}$ as expected. Maclaurin took a different route, one which did not appeal to vanishing quantities. Wishing to prove that $y = nx^{n-1}$, Maclaurin supposed instead that $y = nx^{n-1} + r$ for some positive value r. He argued that this assumption leads to a contradiction of (6.17), which we know is true. Doing the same for the case where r is negative, Maclaurin claimed that r must equal zero.

Such a proof is called a *proof by contradiction*, or (in Latin) a *reductio ad absurdum*. In fact, because Maclaurin considered and eliminated two cases (r positive and r negative), his proof is a 'double' *reductio ad absurdum*. This form of argument was known to the Greeks of Chapter 1, who used it to avoid some of the appeals to the infinitely small that we used there.

We consider the case where, in $y = nx^{n-1} + r$, the value of r is positive. The second inequality in (6.17) becomes

$$nx^{n-1} + r < n(x + h)^{n-1} . \tag{6.18}$$

Maclaurin showed that for any r we choose, we may find a value for h that contradicts (6.18). Consider that

$$nx^{n-1} + r \geq n(x + h)^{n-1} \iff x^{n-1} + \frac{r}{n} \geq (x + h)^{n-1}$$

$$\iff \left(x^{n-1} + \frac{r}{n}\right)^{\frac{1}{n-1}} \geq x + h$$

$$\iff \left(x^{n-1} + \frac{r}{n}\right)^{\frac{1}{n-1}} - x \geq h .$$

Show that there is a positive h satisfying the final inequality no matter what (positive) value r has.

(d) How must we modify the sequence of inequalities at the end of 6.4(c) if r is negative?

6.5 **Newton discovers the generalized binomial theorem.** In a letter intended to reach Leibniz, Newton outlined his discovery of the result that allowed him to expand such expressions as $(a + e)^{1/2}$. His description provides valuable insight into what makes mathematicians tick: the desire to explain patterns.

(a) John Wallis discovered a beautiful expression for π via an ingenious exploration of the area of the circle. Hearing of this, Newton wished to broaden the results. Specifically, Newton noted that

$$x^2 + y^2 = 1 \implies y = (1 - x^2)^{1/2} ,$$

describes a semicircle, so we may profit by studying the entire family of curves of the form $(1 - x^2)^m$. The first several curves in the family appear in Figure 6.11.

When m is an integer, the area beneath these curves between the vertical axis and any $x < 1$ was already known, thanks to results like that in exercise 6.4. Table 1 below summarizes a few cases. Continue the table to its next row where $m = 4$, so that you find an expression for the shaded area in Figure 6.11.

(b) Guess the area for $m = 5$ simply by noticing patterns in Table 1. Explain your thinking in words. Then expand $(1 - x^2)^5$ and so on, to check your

m	$(1-x^2)^m$, expanded	area under $(1-x^2)^m$
0	1	x
1	$1-x^2$	$x - \dfrac{1}{3}x^3$
2	$1-2x^2+x^4$	$x - \dfrac{2}{3}x^3 + \dfrac{1}{5}x^5$
3	$1-3x^2+3x^4-x^6$	$x - \dfrac{3}{3}x^3 + \dfrac{3}{5}x^5 - \dfrac{1}{7}x^7$

Table 1. Areas under the curves in Figure 6.11.

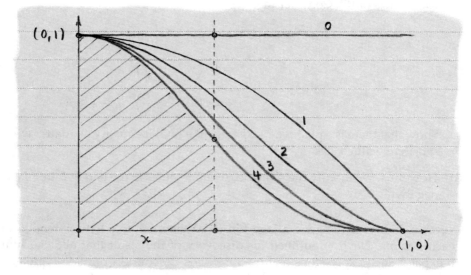

Figure 6.11. Each curve belongs to the family $(1-x^2)^m$ and is labeled with its m value.

work. The goal here is not that your guess is *correct*, but that it is *reasonable*; do not worry if your check fails to match your guess. Rather, use such an experience to modify your guess.

(c) Newton observed that the coefficient of x^3 in each area (in Table 1) is $m \cdot (1/3)x^3$. Puzzling over how he might arrive at the coefficient of x^5 led Newton to guess that he should multiply m by $(m-1)/2$ and multiply that result by $(1/5)x^5$. Check that this is true for $m = 2, 3, 4, 5$.

(d) What should we multiply by $m \cdot (m-1)/2$ in order to determine the value that we should multiply by $(1/7)x^7$? Check your guess.

(e) If the pattern is now clear to you, then use it to discover the area under the circular curve $(1-x^2)^{1/2}$, which was Newton's original interest.

(f) Newton checked his answer to 6.5(e) by squaring it (as best he could, given that the series is infinite) to see if the result was $1 - x^2$. Try this yourself.

6.6 **Euler finds a series involving the exponential constant.** Equation (5.1) equates $\log(1+a)$ with the area under $y = 1/x$ between $x = 1$ and $x = 1+a$. In exercise 5.4, however, we witnessed Henry Briggs choosing the number 10 as the 'base' for his logarithms, via setting $\log 1 = 0$ and $\log 10 = 1$. He chose 10 for its convenience; perhaps the base of the logarithm in (5.1) is a different number.

(a) If $\log 10 = 1$, then (5.1) indicates that the area under $y = 1/x$ between $x = 1$ and $x = 10$ will equal 1. With the help of a figure, argue that this result is impossible.

(b) As we saw in exercise 5.4, Briggs was free to set $\log 10 = 1$, so this is not the trouble. Rather, the base of the logarithm in (5.1) is other than 10. Use the approach you took in 6.6(a) to approximate the number that must replace 10 in $\log 10 = 1$.

(c) The base of a logarithm is traditionally written as a subscript, as in $\log_{10} 10 = 1$. Thanks to the link between logarithms and exponents, we may translate this equality to another: $10^1 = 10$. In general, the equalities

$$\log_b N = L \quad \text{and} \quad b^L = N \tag{6.19}$$

are interchangeable.

Leonhard Euler (Switzerland, born 1707) lived about a century after Briggs, and was primarily responsible for discovering the base of the logarithm in (5.1). The answer is a remarkable number that ranks among the elites of mathematical constants. We follow Euler's lead in this question to see how he captured it.

Let w be an "infinitely small" number that will replace a in (5.1). Thus, the area under $y = 1/x$ between $x = 1$ and $x = 1 + w$ is $\log(1 + w)$. Because w is an infinitesimal, this area is essentially that of a rectangle with height 1 and width w. Thus, Euler sets

$$\log(1 + w) = w \ .$$

Our aim is to pin down the base of this logarithm, an unknown we denote by e for now. Using (6.19), we have

$$\log_e(1 + w) = w \Longrightarrow e^w = 1 + w \ . \tag{6.20}$$

Euler expressed the infinitely small value w as the ratio x/n where x is a constant and n is infinitely *large*. From (6.20), we have

$$(e^w)^n = (1 + w)^n \Longrightarrow e^x = \left(1 + \frac{x}{n}\right)^n \ . \tag{6.21}$$

Euler expanded $(1 + x/n)^n$ using Newton's generalized binomial theorem (as described in exercise 6.5). Produce this expansion yourself until you are satisfied that you can quickly write down subsequent terms.

(d) Finally, Euler claimed that all of the coefficients in the expansion of $(1 + x/n)^n$ equal 1 because n is infinitely large. Why may he claim this?

(e) The upshots of these audacious arguments are that

$$e^x = 1 + x + \frac{x^2}{2!} + \frac{x^3}{3!} + \frac{x^4}{4!} + \cdots \qquad (6.22)$$

and, letting $x = 1$, that

$$e = 1 + 1 + \frac{1}{2!} + \frac{1}{3!} + \frac{1}{4!} + \cdots . \qquad (6.23)$$

Another expression for e comes from (6.21):

$$e = \left(1 + \frac{1}{n}\right)^n \quad \text{where } n \text{ is infinitely large.} \qquad (6.24)$$

The irrational constant $e \approx 2.71828$ is sometimes called the *exponential constant*. The logarithm with e as its base is called *the natural logarithm*, and is written 'ln'.

Euler's discovery of the series (6.22) and the expressions (6.23) and (6.24) perfectly exemplifies his bold approach. At what points during his argument should we be aware that Euler is taking mathematical risks?

7

Notation

Flipping back through the pages of this book, you can see how important geometric figures were in the development of calculus. The figures become more sophisticated as the truths they reveal become deeper; Figure 6.5 of Leibniz, for example, goes to the heart of the connections within calculus, but falls just shy of being an impenetrable maze of lines. Leibniz, as much as anyone in his day, desired to push calculus past the point where its truths are a consequence of diagrams. The notation he invented allowed this, and we use many of his symbols today.

7.1 *Leibniz describes differentials*

The notation of Leibniz underwent a maturing process similar to that of calculus generally. This brief treatment does not attempt to tell the whole story, focusing instead on the final payoff of his efforts.

Leibniz interpreted a curve, like the one in Figure 7.1, as the ratio (at each point) of the curve's vertical motion to its horizontal motion. Mark off equally-spaced divisions on the horizontal, associating each mark (such as A) with the point (B) on the curve directly above it; then mark the corresponding point (C) on the vertical axis. Where the curve has a small vertical rate of change, the points on the vertical axis crowd together.

Leibniz viewed the distances between the marks in Figure 7.1 as *differences* (for example, we may see AD as $OD - OA$) and he chose the notation d to represent such distances. Following d he placed a variable, as in dx. If x represents horizontal distance in Figure 7.1 and y represents vertical distance, then $AD = dx$ and $CF = dy$. The slope of the hypotenuse BE is therefore dy/dx. Leibniz considered these differences "infinitely small" but not zero, making dx incomparably smaller than x; nevertheless, he called dy/dx a ratio of two quantities that are *not* zero. He struggled to reconcile these concepts for his entire life, occasionally resorting to an appeal along the lines of, "This may seem bewildering, but it *works*."

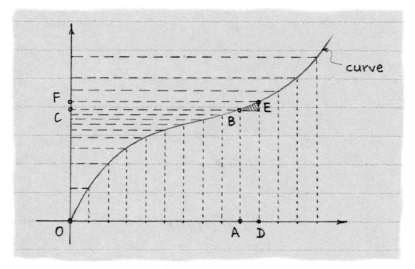

Figure 7.1. A curve, through the eyes of Leibniz, was interpreted as pairs of "infinitely small" differences.

Leibniz coined the word *differential* for infinitely small amounts like dx and dy. This term highlights the notion that dx is a difference; as Figure 7.2 demonstrates for the parabola $y = x^2$, we see that $dx = (x + dx) - x$. In the vertical direction,

$$dy = (x + dx)^2 - x^2$$
$$= x^2 + 2x \cdot dx + (dx)^2 - x^2$$
$$= 2x \cdot dx + (dx)^2 \ . \tag{7.1}$$

When two differentials are multiplied, as in $(dx)^2$, Leibniz claimed that their product was incomparably smaller than either differential. Thus, he discarded the term $(dx)^2$ to conclude that $dy/dx = 2x$. Because dy/dx represents the slope of the tangent line of $y = x^2$ at point A, this result reestablished that of Fermat (in section 3.3) and Cavalieri (in section 4.1).

As further evidence that his claims about differentials were sound, Leibniz produced the differential of a product. If u and v represent two curves, then $d(uv)$ is the differential of their product; we may expand this differential as we did in (7.1) to find that

$$d(uv) = (u + du)(v + dv) - uv$$
$$= uv + u \cdot dv + v \cdot du + du \cdot dv - uv$$
$$= u \cdot dv + v \cdot du \ . \tag{7.2}$$

In the last step, we discarded the product $du \cdot dv$ as incomparably smaller than either du or dv. As a check, we can test $d(x^3) = 3x^2 \cdot dx$ with $u = x$ and $v = x^2$:

$$d(x \cdot x^2) = x \cdot d(x^2) + x^2 \cdot d(x)$$
$$= x \cdot (2x \cdot dx) + x^2 \cdot dx$$
$$= 3x^2 \cdot dx \ . \tag{7.3}$$

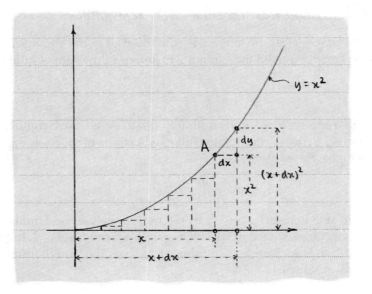

Figure 7.2. Differentials in the context of the parabola $y = x^2$.

Equation (7.2) is the *product rule* for differentials, and it begins to show the power of this notation as a tool for generating results that are difficult to picture. As it happens, it *is* possible to depict (7.2) if we imagine u and v as sides of a rectangle, as in Figure 7.3. Then the area of the rectangle is uv, and $d(uv)$ is an infinitely small increase in that area. This increase is shaded in the figure, and equals the sum of the areas of its three pieces: $u \cdot dv + v \cdot du + du \cdot dv$. The figure, however, does not justify the dismissal of the tiny piece of area $du \cdot dv$ in the corner; after all, does not that tiny piece contribute to the increase in total area? Nevertheless, the accuracy of results like (7.3) lends plausibility to the idea that we may discard $du \cdot dv$.

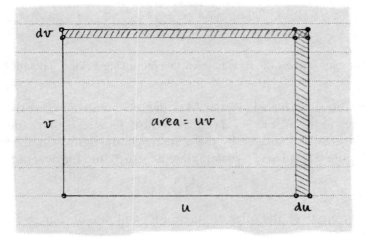

Figure 7.3. This is an attempt at picturing the product rule for differentials.

7.2   The fundamental theorem with new notation

As Figure 7.1 makes clear, distance x is composed of many small segments like $AD = dx$, so Leibniz created the symbol $\int$ to represent the sum of infinitely many lengths. Thus,

$$\int dx = x$$

translates to, "The sum of all of the infinitely small segments of length dx is equal to x." Similarly, $\int dy = y$. This notation allowed Leibniz to concisely express areas as well as lengths; he would write

$$\int x^2 \, dx$$

to refer to the area beneath $y = x^2$. In Figure 7.2, we see the familiar filling of that area with rectangles, each of which has width dx and height x^2. Because dx is infinitely small, so is each area $x^2 \, dx$; thus, $\int x^2 \, dx$ is the sum of infinitely many infinitely small amounts.

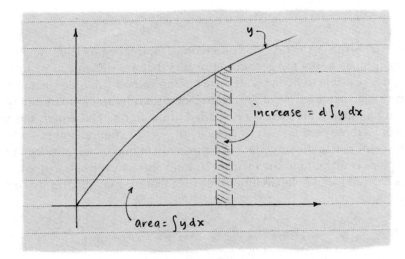

Figure 7.4. We may express the fundamental theorem using Leibniz's notation as shown here.

The quadrature of curve y in Figure 7.4 is identified with the area beneath it, and Leibniz expresses the area as $\int y \, dx$. If we write $d \int y \, dx$, then, we mean a small change in the area, as indicated by the thin strip in the figure. Comparing this to Newton's Figure 6.1, we may translate statement (6.3) into the new notation:

$$y = d \int y \, dx \, . \tag{7.4}$$

As we recall from (6.4), the rules of quadrature and rate of change swap, which translates to

$$y = \int dy \, . \tag{7.5}$$

The inverse nature of quadrature and rate of change appears symbolically when we set the right-hand sides of (7.4) and (7.5) equal to get

$$d \int y\, dx = \int dy \, .$$

(7.6)

There is the fundamental theorem of calculus in a nutshell.

The verb *integrate* entered the mathematical vocabulary around this time, taking the place of *find the quadrature of* as designation for calculating areas. The symbol $\int$ became known as an *integral sign*.

Leibniz used his new notation to describe a rule that we call *integration by parts*. Begin with the product rule for differentials (7.2) and apply infinite summation to both sides:

$$\int d(xy) = \int (x\, dy + y\, dx) \, .$$

(7.7)

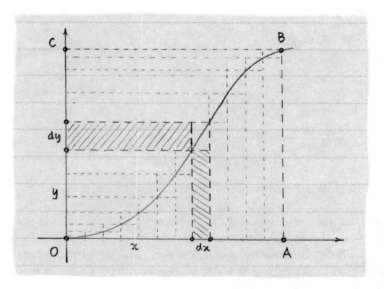

Figure 7.5. We can calculate the area of rectangle $OABC$ as the sum of the two regions on either side of the curve.

The left-hand side is xy by (7.5). Figure 7.5 helps us simplify the right-hand expression. The two shaded rectangles are of area $x\, dy$ and $y\, dx$, so $x\, dy + y\, dx$ is the sum of their areas. Adding all such areas together by writing $\int (x\, dy + y\, dx)$, we get the area of the entire rectangle $OABC$. The rectangle is divided by curve OB into two pieces with area $\int x\, dy$ and $\int y\, dx$. Thus, we may substitute $\int x\, dy + \int y\, dx$ for the right-hand side of (7.7) and rearrange terms to discover

$$\int y\, dx = xy - \int x\, dy \, .$$

(7.8)

Leibniz used this *integration by parts* formula not only to solve a wide variety of new problems, but also to provide new proofs of previously discovered results. In the next section, we will see how Leibniz calculated the area under a cycloid using his formula.

7.3 Leibniz integrates the cycloid

Figure 7.6 re-creates the situation in Figure 4.4, which was used by Roberval to discover his quadrature of the cycloid. In Figure 7.6, we trace the path $AA'G$ of point A from its high point on the circle until it touches the 'ground' at G, thus creating half of a cycloid. Using his notation and his result (7.8), Leibniz confirmed Roberval's conclusion that the area of region $AA'GB$ is 3/2 that of the generating circle with center O.

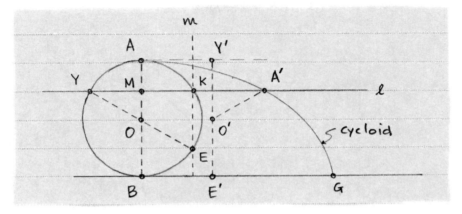

Figure 7.6. The purpose here is simply to argue that arc AK has the same length as KA'.

Desiring the area of the region $AA'GB$ under the cycloid, Leibniz observed that the arc AEB cuts this region into two areas, that of the semicircle AEB and that of the region $AA'GBK$. Because the semicircle occupies half of the generating circle, Leibniz intended to prove that the area of region $AA'GBK$ equals the area of the generating circle.

Construct any line ℓ parallel to BG that intersects AB at M, the generating circle at Y and K, and the cycloid at A'. As the circle rolls to its right, point A travels the path of the cycloid through A', touching ground at G. As the point travels, line ℓ descends, taking M straight from A to B and K around the semicircle from A to B. Segment KA' sweeps through the region of interest $AA'GBK$.

Leibniz first argued that $AK = KA'$, a result that transfers the difficulty of working *outside* the circle to working directly *on* the circle. To this end, draw line m through K parallel to AB, intersecting the circle at E. Imagine the circle rolling to its right; point E will contact BG at some point E'. This will bring diameter YE to position $Y'E'$. At the same time, radius OA will shift to $O'A'$.

The horizontal movement of point Y during this motion is $YM + AY'$, so this is also the horizontal distance traveled by A to A'. By symmetry, we know $YM = MK$ and that the lengths of arcs BE and AK are equal, so the horizontal distance traveled by A is $MK + KA' = YM + KA'$. Thus, $AY' = KA'$ and so

$$\text{the length of arc } AK = \text{the length of } KA' , \tag{7.9}$$

as desired.

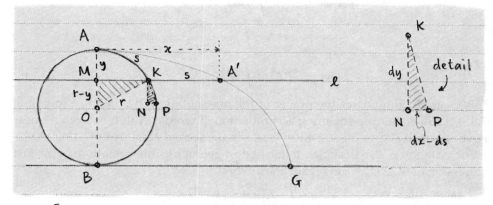

Figure 7.7. Leibniz used characteristic triangle KNP to involve differentials.

With the link between AK and KA' established, we focus on K as it travels from A to B around the generating circle. Figure 7.7 represents the same situation as in Figure 7.6 but with different emphases. For convenience, let $MA' = x$, $AM = y$, $OK = OA = r$, and let the length of arc AK be s. Note that lengths x, y, s are increasing as the circle rolls. With these labels, we introduce the characteristic triangle KNP. Side KN represents the instantaneous vertical change in the position of K as it travels from A to B around the semicircle, so KN equals the differential of $AM = y$. Thus, we may write $KN = dy$ (see the detail in Figure 7.7). In like manner, side KP represents the instantaneous change in the length of arc $AK = s$, so $KP = ds$. Finally, side NP represents the instantanous horizontal change in the position of K. Now (7.9) plays its part: because $MK = MA' - KA' = x - s$, we write $NP = dx - ds$.

Ultimately, we wish to find the area $\int x \, dy$ under the cycloid. Equation (7.8) rearranges to read

$$\int x \, dy = xy - \int y \, dx \tag{7.10}$$

so that the desired area appears on the left-hand side. Leibniz took xy as the area of the rectangle that encloses the region $AA'GB$, so $xy = (\pi r)(2r) = 2\pi r^2$. He found that

$$MK = \sqrt{2ry - y^2}$$

via the Pythagorean Theorem in triangle KMO.

As for $\int y \, dx$, Leibniz calculated dx by noting that the similarity of $\triangle KNP$ and $\triangle KMO$ gives both

$$\frac{ds}{dy} = \frac{r}{MK} = \frac{r}{\sqrt{2ry - y^2}} \tag{7.11}$$

and

$$\frac{dx - ds}{dy} = \frac{r - y}{\sqrt{2ry - y^2}} \Longrightarrow dx = \frac{2r - y}{\sqrt{2ry - y^2}} \, dy \, . \tag{7.12}$$

From (7.10) we conclude that

$$\int x \, dy = xy - \int y \, dx$$

$$= 2\pi r^2 - \int y \, \frac{2r - y}{\sqrt{2ry - y^2}} \, dy$$

$$= 2\pi r^2 - \int \sqrt{2ry - y^2} \, dy \, . \tag{7.13}$$

This example illuminates the power of the new notation. Without it, expressing (7.13) requires an overwhelming accumulation of words. With it, we can harness algebra, which allows for substitutions, as in the second equality of (7.13).

The notation is also designed to facilitate a geometric interpretation. In the last expression of (7.13), we can regard the integral

$$\int \sqrt{2ry - y^2} \, dy$$

as the sum of the areas of the rectangles having length $MK = \sqrt{2ry - y^2}$ and height dy. These rectangles occupy the semicircle ABK, which has area $(1/2)\pi r^2$. Thus, we may continue from (7.13) to conclude that

$$\int x \, dy = 2\pi r^2 - \frac{1}{2}\pi r^2$$

$$= \frac{3}{2}\pi r^2 \, ,$$

confirming Roberval's result.

We might pause and ask how the notation informs us that the rectangles with length MK and height dy occupy the entirety of semicircle ABK; in fact, the notation does *not* inform us, but leaves it up to the context. Later mathematicians, who tweaked integral notation to make it even more readable, had no quarrel with the direction Leibniz took the notation of calculus, adopting it eagerly.

7.4 Furthermore

7.1 **Leibniz sums the reciprocals of the triangular numbers.** Near the beginning of his interest in mathematics, Leibniz moved to Paris and visited **Christiaan Huygens** (Holland, born 1629) to see if he might guide Leibniz in his studies. Leibniz showed Huygens his discoveries about sums and differences, results that later proved conceptually fruitful when Leibniz learned about the mathematics of Fermat and Cavalieri.

Huygens challenged Leibniz to find the sum

$$1 + \frac{1}{3} + \frac{1}{6} + \frac{1}{10} + \frac{1}{15} + \cdots$$

of the reciprocals of the triangular numbers (which were introduced in section 1.4). The implications of the solution Leibniz found trump the solution itself.

(a) Leibniz created a 'harmonic triangle' by writing the terms of the harmonic series in a row and then writing the difference between neighboring terms in the row below. Each succeeding row obeys the same rule. Table 1 shows

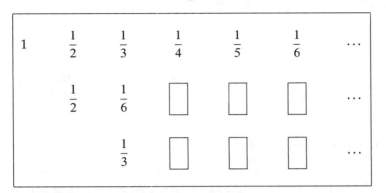

Table 1. The first three rows of the 'harmonic triangle'.

the first few rows, with some terms omitted. Fill in the empty boxes in the table.

(b) Because each term in the second row is the difference of the two neighbors above it, what is the (infinite) sum of the terms in the second row?

(c) How does this result answer the question that Huygens posed to Leibniz?

(d) Similarly, what does this line of reasoning suggest when we apply it to the third row of the table?

   Historical note. Table 1 shares some important features with the one we investigated in exercise 1.2. We use addition to generate rows in one table, and subtraction in the other. We see integers in one table, and reciprocals in the other. Leibniz credited this exercise for the genesis of the idea that differentials (differences) and integration (sums) enjoyed the reciprocal relationship expressed by the fundamental theorem.

7.2 **Leibniz shows how to 'rectify' a curve.** When we *rectify* a problem, we straighten it out, and the same applies to curves. If we wish to know the length of curve OBE in Figure 7.1, for example, we can lay a string over it and then pull the string taut and measure it. Leibniz used his new notation to mathematically accomplish this task, which modern scholars call 'finding an *arc length*.'

(a) The shaded characteristic triangle in Figure 7.1 has hypotenuse BE, which we label ds. This differential approximates the tiny bit of curve between B and E. Label the side of the characteristic triangle that is equal to AD as dx and the side that is equal to CF as dy. Use the Pythagorean Theorem and a bit of algebra to show that

$$ds = \sqrt{1 + \left(\frac{dy}{dx}\right)^2}\, dx\ .$$

(b) Leibniz calculated the length of curve AE by summing the differentials ds, concluding that the expression

$$\int ds = \int \sqrt{1 + \left(\frac{dy}{dx}\right)^2}\, dx$$

rectifies curve AE. Thus, he changed a curve length problem into one of quadrature. Try this approach on the curve $y = x$ from $x = 0$ to $x = 1$, and check your answer without calculus.

7.3 Leibniz illustrates the fundamental theorem of calculus.

Chapter 6 would have been an appropriate setting for Figure 7.8, which is similar to the figure that Leibniz used to illustrate the fundamental theorem of calculus. Placing the figure in this chapter, however, allows us to analyze it using his advantageous notation.

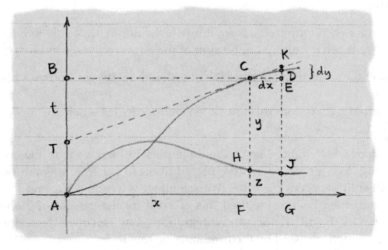

Figure 7.8. Leibniz used a figure like this one to establish the inverse relationship between quadrature and rates of change.

Suppose curve AHJ passes through the origin as shown. Let F and G be the points along the horizontal axis that correspond to H and J. We will construct curve ACD as a companion to curve AHJ so that the following is always true: if we draw a tangent line to curve ACD through C that intersects the vertical axis at T, and create triangle TBC with BC parallel to the horizontal axis AF, then we require that

$$BT = AF \cdot FH. \tag{7.14}$$

If we draw curve ACD according to this stipulation, then (Leibniz claims) we can show that FC equals the area of the region $AHFA$.

Following Leibniz's lead, we let $AF = x$, $FC = y$, $FH = z$, and $TB = t$. Infinitely small extensions of TC to K and BC to E create characteristic triangle CEK similar to triangle CBT. Because CE and EK are small changes in x and y respectively, we can label these sides dx and dy as shown. So $t/x = dy/dx$ by similar triangles, and $t = xz$ by (7.14). These two observations together yield

$$z = \frac{dy}{dx}, \tag{7.15}$$

so curve AHJ describes the rate of change of curve ACD.

From (7.15) we get $z\, dx = dy$ and, using the ideas and notation of section 7.2, we have

$$\int z\, dx = \int dy = y.$$

Now y is the vertical part of curve ACD at F and Leibniz interprets $\int z\, dx$ as the sum of all of the infinitely thin rectangles with height z and width dx between points A and F. In other words, Leibniz equates $\int z\, dx$ with the area of region $AHFA$.

(a) With this in mind, explain the visual relationship between curves ACD and AHJ. In particular, explain why the point where ACD is "steepest" corresponds to the maximum point on AHJ.

(b) Which of Newton's observations (6.3) or (6.4) has Leibniz established by showing that $y = \int z\, dx$?

Historical note: A curve that has its vertical part equal to the area beneath another curve is called a *quadratrix*. In Figure 7.8, then, curve ACD is the quadratrix of curve AHJ.

7.4 **Euler explains the product and quotient rules.** Euler followed Leibniz in claiming that a differential represented an infinitely small increment in a variable, so that $x + dx$ and x are equal. Further, any product of differentials, such as $dx \cdot dy$ or $(dx)^3$, diminished the infinite smallness yet again, as though such products were 'infinitely infinitely small'.

These claims underlie Euler's explanation of how we may find the differential of the product of two expressions. Take $y = x^2(1 - x)^{1/2}$ as an example; we may not algebraically combine the factors x^2 and $(1 - x)^{1/2}$, so we take Euler's approach and label the factors $u = x^2$ and $v = (1 - x)^{1/2}$. Now we rephrase our problem: if $y = u \cdot v$, then what is dy?

Euler argued that

$$y = u \cdot v \implies y + dy = (u + du)(v + dv)$$
$$\implies dy = uv + u \cdot dv + v \cdot du + du \cdot dv - uv$$
$$\implies dy = u \cdot dv + v \cdot du ,$$

a result known as the *product rule* for differentials, which we met before as (7.2). For our specific example above, we have the tools to calculate $du = 2x \cdot dx$, but we cannot cope with dv just yet. In exercise 7.6, we address this.

The rest of this question sketches Euler's development of a *quotient rule* for differentials that answers the question: if $y = u/v$, then what is dy?

Expressing $y = u/v$ as $y = uv^{-1}$ allows us to use the product rule to find that

$$dy = u \cdot d(v^{-1}) + v^{-1} \cdot du . \tag{7.16}$$

Euler approached the differential $d(v^{-1})$ in the same way he began deriving the product rule. Let $q = 1/v$, so

$$q + dq = \frac{1}{v + dv} = \frac{1}{v} \left(\frac{1}{1 + dv/v} \right) .$$

His clever factoring results in an expression involving the sum of a geometric series; see (2.7). Swap in the appropriate geometric series, and simplify as Euler would until you reach the result

$$dy = \frac{v \cdot du - u \cdot dv}{v^2} .$$

7.5 **Euler calculates the differentials of $\ln x$ and e^x.** When e is the base of a logarithm, we write $\ln x$ (as described in exercise 6.6). Euler derived the differential of $y = \ln x$ using the same assumptions as those described in exercise 7.4.

Thanks to the links we have drawn between $y = \ln x$ and the hyperbola $y = 1/x$, we suspect that

$$d(\ln x) = \frac{1}{x} dx .$$

Here we see Euler's explanation of this result.

(a) He began with

$$y = \ln x \implies y + dy = \ln(x + dx)$$
$$\implies dy = \ln \left[(x) \left(1 + \frac{dx}{x} \right) \right] - \ln x .$$

His unusual factoring shows its value when we apply (5.12) and (5.10) in turn. Eliminating products of differentials completes the task. Write down the details of this process.

(b) Use the same approach to find dy when $y = e^x$, and discover the intriguing differential

$$d(e^x) = e^x \, dx \ .$$

7.6 **The chain rule.** We required the differential $d((1 - x)^{1/2})$ in exercise 7.4, but lacked a simple rule for calculating it. We remedy that now. Suppose we let $y = (1 - x)^{1/2}$ and $u = 1 - x$, so that $y = u^{1/2}$. Using Euler's approach as in exercise 7.4, we have

$$y + dy = (u + du)^{1/2} \implies dy = (u + du)^{1/2} - u^{1/2} \ .$$

Expand $(u + du)^{1/2}$ using the generalized binomial theorem, so

$$dy = \left(u^{1/2} + \frac{1}{2} u^{-1/2} du - \frac{1}{8} u^{-3/2} (du)^2 + \cdots \right) - u^{1/2} \ .$$

We drop all terms that include a product of differentials to see that

$$dy = \frac{1}{2} u^{-1/2} \, du \ .$$

Because $du = d(1 - x) = -dx$, we conclude that

$$dy = \frac{1}{2} (1 - x)^{-1/2} (-dx) = -\frac{1}{2} (1 - x)^{-1/2} dx \ .$$

(a) In search of a more general rule, we can step back through the argument with a variable like m in place of the exponent, so that our goal is to find dy when $y = (1 - x)^m$. Try this.

(b) Now try with u in place of $1 - x$, to see if the argument remains valid no matter what expression we substitute for u in $y = u^m$.

(c) The rule we seek is commonly called the *chain rule* for its use in situations where a variable is expressed in terms of another, which is itself expressed in terms of a third. (The 'links' from one variable to the next create a 'chain'.) For example, if $y = u^{1/2}$ and $u = 1 - x$, we can return to the expression $y = (1 - x)^{1/2}$ studied earlier in this question.

Similarly, we can chain together $y = \ln u$ and $u = 2x$ to get $y = \ln(2x)$, or $y = u^{-1}$ and $u = 1 + 2x$ to get $y = (1 + 2x)^{-1}$, or $y = e^u$ and $u = -x^2$ to get $y = e^{-x^2}$. For each of these three expressions, explain how we may reach the conclusions

$$y = \ln(2x) \implies dy = \frac{1}{x} \, dx,$$

$$y = (1 + 2x)^{-1} \implies dy = -2(1 + 2x)^{-2} \, dx,$$

$$y = e^{-x^2} \implies dy = -2x e^{-x^2} \, dx.$$

8

Chords

"Where is that?" is a question as ancient as astronomy, often accompanied by, "Where is it going?" and, these days, "Is that thing going to hit us?" Because Greek thinkers of old believed that the earth was stationary and that celestial objects traveled in circular paths, the study of angles related to circles received careful attention. The word for this study, *trigonometry*, refers to the measure of triangles, which yield a multitude of curious and beautiful truths.

Greek astronomers were privy to many such truths, but it was Indian scholars in the years between 400 and 700 who began to link the measure of angles to series. The mathematicians we studied in Chapter 2 were then able to complete this project, stopping just shy of results that might have led us to call them the discoverers of calculus. Because their arguments are somewhat sophisticated, they have been delayed to this point, where we can make use of the notation of Leibniz, and where readers should be thoroughly warmed up to the task.

8.1 Preliminary results known to the Greeks

Triangles, circles, and angles appear in Figure 8.1, a simple picture of Earth at the center of a circular orbit. Some celestial object moves from A along the arc to B. Because $\triangle EGD$ is similar to $\triangle EFB$, the ratios of each side to each of the others is constant, no matter the lengths of the sides. Thus, for each angle like $\angle DEG$ we may calculate these ratios once and for all. Greek astronomers were able to do this with the help of ingenious formulas that link these ratios to one another.

Our names for these ratios have a curious origin. Because arc APB and segment AFB in Figure 8.1 look like a bow with its string, Indian scholars used the word *ardha-jyā*, meaning *half-bowstring*, to refer to FB. The abbreviated term *jyā* became *jiba* as these studies filtered from India to Islamic universities. Vowels are often omitted in Arabic; a Latin translator misread *jiba* as *jaib*, Arabic for *bay*, so wrote *sinus* in his translation. Thus do we use the words *sine* to refer to FB and *cosine* for EF.

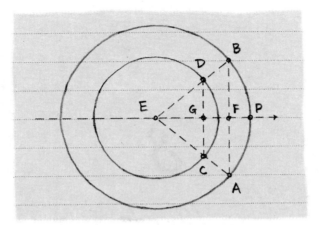

Figure 8.1. Similar triangles have proportional sides, and these constant proportions give rise to trigonometric relationships.

One more result from the Greeks before we focus on India. A *central angle*, like $\angle PEB$ in Figure 8.1, has its vertex at the center of a circle. The Greeks equated central angles with the length of the arc that the angle subtends on a unit circle. In particular, if $EB = 1$, then the measure of $\angle PEB$ equals the length of arc PB. We say now that angles measured in this way are in *radians* rather than *degrees*. Because the entire circumference of a unit circle is 2π, the central angle of the circle as a whole is 2π radians.

A circle's circumference ought to be proportional in length to the entire central angle in the same ratio as any particular arc length is to the angle corresponding to that arc. In symbolic form, the previous sentence states that

$$\frac{2\pi \cdot EB}{2\pi} = \frac{PB}{\angle PEB} \implies PB = \angle PEB \cdot EB$$

for the particular arc PB. More generally, if r is the radius of a circle and the central angle α intercepts an arc of length s, then

$$s = \alpha \cdot r .$$

We will find these few facts about angles, triangles, and circles useful as we fast-forward a few centuries to India.

8.2   Jyesthadeva finds series for sine and cosine

As was mentioned in Chapter 2, we do not know which arguments authored by Jyesthadeva in the early 1500s belong to him and which belong to others, like Nilakantha. Bear this in mind as we attribute all results here to Jyesthadeva, for the sake of simplicity. These involve arguments so familiar to us by now that the notation developed by Leibniz fits naturally, and we will use it. However, the following proof is unique among those encountered thus far because what emerges is not one truth, but two. It is easy to imagine that this proof evolved via a good deal of refining on the part of Indian scholars.

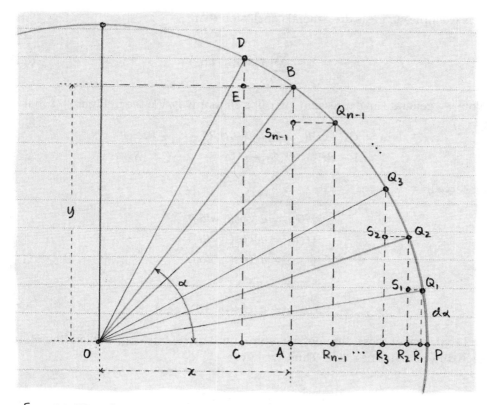

Figure 8.2. We seek expressions for the lengths OA and AB in $\triangle OAB$, where $OB = 1$.

The dual punchlines are a pair of series, one equal to the sine and the other equal to the cosine of any angle. In the unit circle shown in Figure 8.2, the sine of $\angle POB = \alpha$ is the length of AB. Jyesthadeva treated α as the sum of n tiny angles $d\alpha$. This treatment creates n triangles with central angles $d\alpha, 2 \cdot d\alpha, 3 \cdot d\alpha, \ldots$ that culminate in $\triangle OAB$ with central angle α.

Let $OA = x$ and $AB = y$ so that we may label the characteristic triangle DEB with the differentials $d\alpha, dx$, and dy. Because $x = \cos\alpha$ and $y = \sin\alpha$, these two variables are our ultimate focus.

Figure 8.2 shows $\triangle OR_1Q_1$, $\triangle OR_2Q_2$, and $\triangle OR_3Q_3$ created by angles $d\alpha$, $2 \cdot d\alpha$, and $3 \cdot d\alpha$ respectively. Subsequent triangles are not shown until we reach $\triangle OR_{n-1}Q_{n-1}$ and, finally, $\triangle OAB$. Although it is true that

$$y = R_1Q_1 + S_1Q_2 + S_2Q_3 + \cdots + S_{n-1}B ,$$

we can improve this awkward expression. Because $d\alpha$ is a differential, each of the lengths in the expression is also a differential. Of what distance is S_2Q_3 (for example) a differential? The length of $S_2 Q_3$ is a small vertical change in the length of $R_2 Q_2 = \sin(2 \cdot d\alpha)$, so we may write

$$S_2 Q_3 = d\left(\sin(2 \cdot d\alpha)\right) .$$

This example suggests the general conclusion that

$$y = \sum_{k=0}^{n-1} d\left(\sin(k \cdot d\alpha)\right) . \tag{8.1}$$

Now we generate an expression for x in a similar way. We see in Figure 8.2 that

$$x = OP - PR_1 - R_1 R_2 - R_2 R_3 - \cdots - R_{n-1} A$$
$$= 1 - PR_1 - Q_1 S_1 - Q_2 S_2 - \cdots - Q_{n-1} S_{n-1}.$$

Observe that

$$PR_1 = d\left(\cos(0 \cdot d\alpha)\right),$$
$$Q_1 S_1 = d\left(\cos(1 \cdot d\alpha)\right),$$
$$Q_2 S_2 = d\left(\cos(2 \cdot d\alpha)\right),$$
$$\vdots$$
$$Q_{n-1} S_{n-1} = d\left(\cos((n-1) \cdot d\alpha)\right),$$

and that each of these differentials represents a move from right to left; therefore, each differential is negative. Thus, we have

$$x = 1 + \sum_{k=0}^{n-1} d\left(\cos(k \cdot d\alpha)\right) . \tag{8.2}$$

Jyesthadeva linked (8.1) and (8.2) by relating the differentials of sines to cosines and vice versa. For clarity, we use Figure 8.3 rather than the busy Figure 8.2. Radius OB meets differential BD at a right angle, so $\triangle DEB$ is similar to $\triangle OAB$. Thus,

$$\frac{DE}{BD} = \frac{OA}{OB} \implies \frac{d(\sin\alpha)}{d\alpha} = \frac{\cos\alpha}{1}$$
$$\implies d(\sin\alpha) = \cos\alpha \cdot d\alpha . \tag{8.3}$$

To take a specific case from Figure 8.2, we translate (8.3) to see that

$$S_2 Q_3 = OR_2 \cdot d(2 \cdot d\alpha) .$$

We may also deduce the analogue of (8.3) for cosine:

$$\frac{BE}{BD} = \frac{AB}{OB} \implies d(\cos\alpha) = -\sin\alpha \cdot d\alpha . \tag{8.4}$$

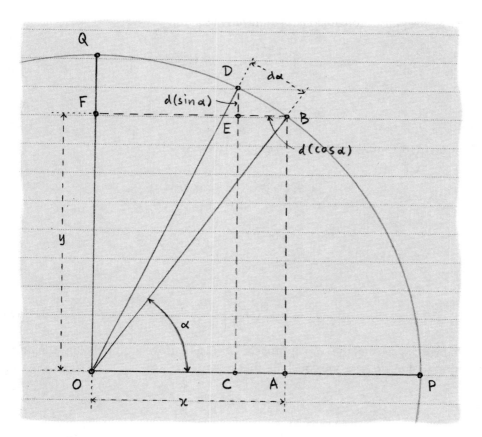

Figure 8.3. Differentials ED and EB link to the sine and cosine of α.

Now in Jyesthadeva's toolkit, these formulas unlock series for $x = \cos\alpha$ and $y = \sin\alpha$ via a clever (and infinite) give and take. Substituting $k \cdot d\alpha$ for α in (8.4), we may alter (8.2) to remove the differential of cosine:

$$x = 1 + \sum_{k=0}^{n-1} d(\cos(k \cdot d\alpha))$$

$$= 1 + \sum_{k=0}^{n-1} \left[-\sin(k \cdot d\alpha) \cdot d(k \cdot d\alpha) \right] . \tag{8.5}$$

Jyesthadeva simplified (8.5) thanks to the following two observations about k and $d\alpha$. First, the differential $d(k \cdot d\alpha)$ equals the difference between angles $(k + 1) \cdot d\alpha$ and $k \cdot d\alpha$, and this difference is always $d\alpha$. Second, the quantity $d\alpha$ is infinitely small, and for small angles like $k \cdot d\alpha$ we may use the approximation

$$\sin(k \cdot d\alpha) \approx k \cdot d\alpha . \tag{8.6}$$

Figure 8.3 justifies this claim; observe that for small α, the length of $AB = \sin\alpha$ comes close to the length of the arc $BP = \alpha$.

Using this pair of conclusions, we simplify (8.5) to

$$x \approx 1 + \sum_{k=0}^{n-1} \left(-(k \cdot d\alpha) \cdot d\alpha \right)$$

$$= 1 - (d\alpha)^2 \sum_{k=0}^{n-1} k \ . \tag{8.7}$$

The sum in (8.7) may, with the help of (1.10), be rewritten

$$\sum_{k=0}^{n-1} k = 1 + 2 + 3 + \cdots + (n-1)$$

$$= \frac{1}{2}(n-1)n$$

$$= \frac{n^2}{2} - \frac{n}{2}$$

$$= \frac{n^2}{2} \left[1 - \frac{1}{n} \right] , \tag{8.8}$$

so that we can see that the sum approaches $n^2/2$ as n grows infinitely large. Bearing in mind that $n \cdot d\alpha = \alpha$, we conclude that

$$\cos\alpha \approx 1 - (d\alpha)^2 \cdot \frac{n^2}{2}$$

$$= 1 - \frac{\alpha^2}{2} \ . \tag{8.9}$$

We may rightly view (8.9) as a place to catch our breath. Figure 8.4 shows the two curves in (8.9), indicating that the approximation is most accurate when α is near zero; this makes sense in the context of (8.6) where the approximation entered the argument. Any uneasiness we felt with (8.6) suggests itself in Figure 8.3: the further α is from zero, the worse (8.6) does as an approximation. To improve it, Jyesthadeva detoured through an approximation for $\sin\alpha$ that not only makes use of (8.9) but even sharpens it.

We now do for $y = \sin\alpha$ what we earlier did for $x = \cos\alpha$ by using (8.3) in (8.1) to see that

$$y = \sum_{k=0}^{n-1} \left(\cos(k \cdot d\alpha) \cdot d(k \cdot d\alpha) \right) . \tag{8.10}$$

We re-use our observation that $d(k \cdot d\alpha) = d\alpha$ and apply (8.9) to discover that

$$y \approx \sum_{k=0}^{n-1} \left[\left(1 - \frac{(k \cdot d\alpha)^2}{2} \right) \cdot d\alpha \right]$$

$$= n \cdot d\alpha - \frac{(d\alpha)^3}{2} \sum_{k=0}^{n-1} k^2 \ . \tag{8.11}$$

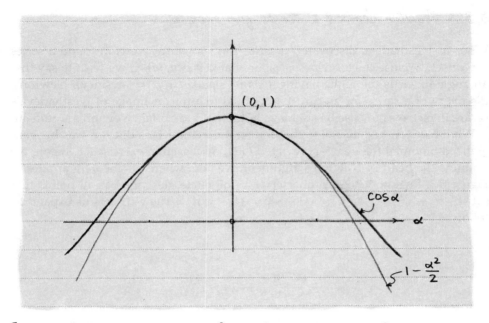

Figure 8.4. This sketch of $\cos \alpha$ and $1 - \alpha^2/2$ shows the close approximation when α is near zero.

Algebra similar to (8.8) shows that the sum of squares in (8.11) approaches $n^3/3$ as n grows infinitely large, so

$$\sin \alpha \approx \alpha - \frac{\alpha^3}{2 \cdot 3} \, . \tag{8.12}$$

We may catch our breath again, and then go back to (8.5) armed with (8.12) to show that

$$\cos \alpha \approx 1 - \frac{\alpha^2}{2} + \frac{\alpha^4}{2 \cdot 3 \cdot 4}$$

in a line of reasoning that involves the sum of cubes. Jyesthadeva dodged back and forth between the approximations for $\sin \alpha$ and $\cos \alpha$ until he was satisfied that

$$\sin \alpha = \alpha - \frac{\alpha^3}{3!} + \frac{\alpha^5}{5!} - \frac{\alpha^7}{7!} + \cdots \tag{8.13}$$

and

$$\cos \alpha = 1 - \frac{\alpha^2}{2!} + \frac{\alpha^4}{4!} - \frac{\alpha^6}{6!} + \cdots \, . \tag{8.14}$$

These wonderful expansions are just one of the many reasons why Newton put so much stake in series. He derived them himself from scratch, over a hundred years later; his path took a twist that Indian mathematicians did not anticipate.

8.3   *Newton derives a series for arcsine*

Whereas Jyesthadeva discovered a series for $y = \sin \alpha$, Newton found a series for α itself. We use the term *arcsine*, abbreviated *arcsin*, when we wish to say that an angle like α is the angle having y as its sine, as in $\alpha = \arcsin(y)$. Newton's discovery of a series for $\arcsin(y)$ amounted to a mirror image of Jyesthadeva's series. In fact, we may use Jyesthadeva's Figure 8.3 to explain Newton's argument.

What Newton had available that Indian mathematicians did not was his generalized binomial theorem, abbreviated GBT, discussed in exercise 6.5. Before we examine the heart of Newton's argument, we note where his theorem applies in Figure 8.3. Region OPQ is a quarter of a unit circle, described as all points that satisfy $x^2 + y^2 = 1$ or $x = \sqrt{1 - y^2} = (1 - y^2)^{1/2}$. The GBT lets us expand as follows:

$$(1 - y^2)^{1/2} = 1 + \frac{1/2}{1}(-y^2)$$

$$+ \frac{(1/2)(-1/2)}{1 \cdot 2}(-y^2)^2$$

$$+ \frac{(1/2)(-1/2)(-3/2)}{1 \cdot 2 \cdot 3}(-y^2)^3$$

$$+ \frac{(1/2)(-1/2)(-3/2)(-5/2)}{1 \cdot 2 \cdot 3 \cdot 4}(-y^2)^4 + \cdots$$

$$= 1 - \frac{1}{2}y^2 - \frac{1}{8}y^4 - \frac{1}{16}y^6 - \frac{5}{128}y^8 - \cdots . \tag{8.15}$$

Although this is not exactly how Newton used the GBT in his argument, this expansion serves as a reminder as to how it works. You might try graphing the first few terms on the right-hand side of (8.15) to see how well they approximate a quarter circle.

Looking back at Figure 8.3, we observe that the similar pair $\triangle BDE$ and $\triangle OBF$ gives us

$$\frac{BD}{DE} = \frac{OB}{BF} \implies \frac{d\alpha}{dy} = \frac{1}{x}$$

$$\implies d\alpha = \frac{dy}{x} = (1 - y^2)^{-1/2} dy .$$

If we expand $(1 - y^2)^{-1/2}$ using the GBT, we find that

$$d\alpha = \left(1 + \frac{1}{2}y^2 + \frac{3}{8}y^4 + \frac{5}{16}y^6 + \frac{35}{128}y^8 + \frac{63}{256}y^{10} + \cdots\right) dy .$$

This relationship between $d\alpha$ and dy could hardly be guessed from Figure 8.3. We may sum the differentials $\int d\alpha = \alpha = \arcsin y$ and use our observations in section 7.2 to find

$$\arcsin y = \int \left(1 + \frac{1}{2}y^2 + \frac{3}{8}y^4 + \frac{5}{16}y^6 + \frac{35}{128}y^8 + \frac{63}{256}y^{10} + \cdots \right) dy$$

$$= \int dy + \int \frac{1}{2}y^2 \, dy + \int \frac{3}{8}y^4 \, dy + \int \frac{5}{16}y^6 \, dy + \cdots$$

$$= y + \frac{1}{6}y^3 + \frac{3}{40}y^5 + \frac{5}{112}y^7 + \frac{35}{1152}y^9 + \frac{63}{2816}y^{11} + \cdots .$$

The fractions in this sum lack the appeal of, say, those in (8.13) and (8.14), Jyesthadeva's series for sine and cosine. Nothing about the sequence of numerators $1, 3, 5, 35, \ldots$ or denominators $6, 40, 112, 1152, \ldots$ immediately suggests an elegant pattern. A pattern is there, however, hidden in the relationships between the numerators and denominators.

When faced with a sequence where a pattern might hide, mathematicians often try factoring. In the factored sequence of fractions

$$\frac{1}{2 \cdot 3}, \quad \frac{3}{2^3 \cdot 5}, \quad \frac{5}{2^4 \cdot 7}, \quad \frac{5 \cdot 7}{2^7 \cdot 9}, \quad \frac{7 \cdot 9}{2^8 \cdot 11}, \tag{8.16}$$

we see the importance of odd integers and a curious increase in the number of 2's in the denominator. The exponents $1, 3, 4, 7, 8, \ldots$ might trigger a thought in a pattern-seeker: these are the numbers of 2's in the corresponding sequence

$$2 = 2,$$
$$2 \cdot 4 = 2^3,$$
$$2 \cdot 4 \cdot 6 = 2^4 \cdot 3,$$
$$2 \cdot 4 \cdot 6 \cdot 8 = 2^7 \cdot 3,$$
$$2 \cdot 4 \cdot 6 \cdot 8 \cdot 10 = 2^8 \cdot 3 \cdot 5,$$

$$\cdots$$

This sequence smacks of factorials, as do the odd integers in (8.16). From here it is a matter of experimentation to ultimately discover that

$$\arcsin y = y + \frac{1^2}{3!} \, y^3 + \frac{1^2 \cdot 3^2}{5!} \, y^5 + \frac{1^2 \cdot 3^2 \cdot 5^2}{7!} \, y^7 + \cdots$$

$$= y + \frac{1}{2} \cdot \frac{y^3}{3} + \frac{1 \cdot 3}{2 \cdot 4} \cdot \frac{y^5}{5} + \frac{1 \cdot 3 \cdot 5}{2 \cdot 4 \cdot 6} \cdot \frac{y^7}{7} + \cdots .$$

Now it is no trouble to generate more terms in the series.

In order to trust the pattern, of course, we must first prove it to be correct; then, to trust the result, we must find the values of y that cause the series to speak a truth. But once these (non-trivial) matters are addressed, we might be tempted to say that Newton's discovery has unlocked the mystery of arcsine. Mathematicians would disagree, asking: Why do odd and even integers play such important roles, when arcsine concerns angles and circles? Might another way to describe arcsine make more sense? Until questions like these are answered, mathematicians press on.

Sometimes, curiosity is driven by criticism. The successes we have witnessed in this book were borne on the shoulders of some questionable assumptions about the infinite. Subsequent to the publication of the calculus by Newton and Leibniz, other thinkers rightly criticized their assumptions. The next chapter tells their story.

8.4   Furthermore

8.1 **The reciprocals of the squares.** Sometimes a problem sounds quite simple, but takes a surprisingly long time to solve. Such a problem was to find the sum

$$\frac{1}{1\cdot 1} + \frac{1}{2\cdot 2} + \frac{1}{3\cdot 3} + \frac{1}{4\cdot 4} + \cdots \tag{8.17}$$

of the 'reciprocals of the squares.' Stated in the mid-1600s, this was not evaluated for eighty years, despite the efforts of many. **Johann Bernoulli** (Switzerland, born 1667) compared it to the sum

$$1 + \frac{1}{1\cdot 2} + \frac{1}{2\cdot 3} + \frac{1}{3\cdot 4} + \cdots \tag{8.18}$$

with the purpose of proving that (8.17) converges.

(a) Prove that (8.18) converges to 2 by expressing each fraction as the difference of a pair of fractions. If you choose wisely, each pair of fractions of (8.18) will simplify with the help of its neighbors.

(b) Why does the fact that (8.18) converges have any bearing on whether (8.17) does as well?

(c) While we are studying (8.18), we pause to see Bernoulli's proof that the harmonic series (3.12) diverges. Dropping the initial 1 from the harmonic series, Bernoulli wrote the remaining terms first as

$$\frac{1}{2} + \frac{2}{6} + \frac{3}{12} + \frac{4}{20} + \frac{5}{30} + \frac{6}{42} + \cdots$$

and then as

$$\frac{1}{2} + \frac{1}{6} + \frac{1}{12} + \frac{1}{20} + \frac{1}{30} + \frac{1}{42} + \cdots$$

$$\frac{1}{6} + \frac{1}{12} + \frac{1}{20} + \frac{1}{30} + \frac{1}{42} + \cdots$$

$$\frac{1}{12} + \frac{1}{20} + \frac{1}{30} + \frac{1}{42} + \cdots$$

$$\frac{1}{20} + \frac{1}{30} + \frac{1}{42} + \cdots$$

$$\vdots$$

He summed each row, starting at the top, using each sum to help determine the next. How does this process lead to the desired result?

8.2 **Euler sums the reciprocals of the squares.** Euler first approximated (8.17) and then discovered the exact solution. This question steps through the latter argument.

(a) First, a detour through the finite: the equation

$$(x - a)(x - b) = 0$$

has the two solutions $x = a$ and $x = b$. Expanding and dividing by ab transforms this equation to

$$\frac{1}{ab}x^2 - \left(\frac{1}{a} + \frac{1}{b}\right)x + 1 = 0,$$

where we observe that the constant term is 1, and the coefficient of x is the sum of the reciprocals of the solutions, with the sign changed. Check that the same is true if we start with

$$(x - a)(x - b)(x - c) = 0.$$

(b) Euler guessed that this result might hold in situations where the equation has an infinite number of solutions. He began with $\sin\alpha = 0$, which has solutions at all multiples of π. Euler replaced $\sin\alpha$ with its series in (8.13); dividing by α and substituting u for α^2 gives

$$1 - \frac{u}{3!} + \frac{u^2}{5!} - \frac{u^3}{7!} + \cdots = 0,$$

which is (an infinite) polynomial with 1 as its constant. How do the results from 8.2(a) apply to allow Euler to conclude that (8.17) sums to $\pi^2/6$?

(c) The identity

$$\frac{\pi^2}{8} = \sum_{k=1}^{\infty} \frac{1}{(2k - 1)^2}$$

closely resembles the one that we pursued in 8.2(b), and can be proven using (8.14). An alternate proof relies simply on a bit of term sorting. Convince yourself by one of these methods (or by one of your own) that the identity for $\pi^2/8$ is correct.

8.3 **Euler's differential of $\sin\alpha$.** Euler's treatment of the differential of $\sin\alpha$ featured his characteristic mix of brevity, clarity, and audacity. His aim was to prove (8.3), as Jyesthadeva did; unlike Jyesthadeva, however, Euler did not appeal directly to geometry.

Euler began by substituting $d\alpha$ for α in both (8.13) and (8.14):

$$\sin(d\alpha) = d\alpha - \frac{(d\alpha)^3}{3!} + \frac{(d\alpha)^5}{5!} - \frac{(d\alpha)^7}{7!} + \cdots,$$

$$\cos(d\alpha) = 1 - \frac{(d\alpha)^2}{2!} + \frac{(d\alpha)^4}{4!} - \frac{(d\alpha)^6}{6!} + \cdots.$$

(*Historical note.* These series were known to Euler not via the work of Jyestha-deva, but thanks to the efforts of Newton. In section 8.3, we studied Newton's discovery of a series for arcsin α. Newton unraveled this result to subsequently discover the series (8.13) for sin α.)

As did Leibniz, Euler treated any power (greater than 1) of a differential like $d\alpha$ as incomparably smaller than $d\alpha$ itself. Thus, $\sin(d\alpha) = d\alpha$ and $\cos(d\alpha) = 1$. In the same manner, Euler treated the quantities like α and $\alpha + d\alpha$ as equal, so

$$y = \sin\alpha \implies y + dy = \sin(\alpha + d\alpha) \, .$$

Using the trigonometric identity

$$\sin(A + B) = \sin A \cos B + \cos A \sin B \, ,$$

Euler concluded that

$$y + dy = \sin\alpha \cos(d\alpha) + \cos\alpha \sin(d\alpha)$$
$$= \sin\alpha + \cos\alpha \cdot d\alpha \, .$$

Subtracting $y = \sin\alpha$ from both sides, Euler arrived at his goal:

$$d(\sin\alpha) = \cos\alpha \cdot d\alpha \, .$$

This conclusion matches that of Jyesthadeva in section 8.2. Argue as Euler would to find the differential of $\cos\alpha$.

8.4 **Pascal's 'sum of sines' in the notation of Leibniz.** In exercise 6.1, we outlined an argument by Pascal that had at its core the link between tangency and quadrature that is the hallmark of the fundamental theorem. Now that we have new notation, and some familiarity with trigonometry, we may translate Pascal's result.

Figure 8.5 re-creates some of Figure 6.7, shifting emphasis to the angles that correspond to the vertical segments. Here, we let $\alpha = \angle DOB$, $\beta = \angle VOB$, and $\theta = \angle UOB$. Assume the quarter-circle has radius 1 for now. With these labels, explain why

$$\int \sin\alpha \, d\alpha = \cos\beta - \cos\theta$$

correctly restates Pascal's conclusion in exercise 6.1. (In the infinite sum, the angle α ranges from β to θ.)

Historical note. Later mathematicians simplified the notation further by placing the lower and upper bounds for α on the summation symbol itself, like so:

$$\int_{\beta}^{\theta} \sin\alpha \, d\alpha = \cos\beta - \cos\theta \, .$$

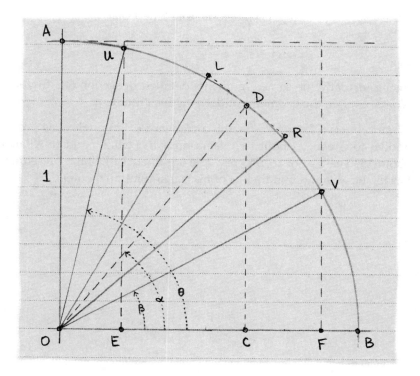

Figure 8.5. This version of Figure 6.7 emphasizes the angles.

8.5 **Leibniz (re)discovers a series for** arctan z. Using Figure 6.6, we observed how Leibniz rediscovered a series equal to $\pi/4$. The same figure can lead us to a series for the arctangent of an angle.

(a) Let $\angle ADB = 2\theta$ (the 2 is for convenience only). Explain these four facts:
  - that $\angle ABE = \theta$,
  - that the area of sector ADB is θ,
  - that $z = x/y$,
  - and that $y = z(2 - x)$.

(b) Because $\tan\theta = x/y$, we have

$$\arctan z = \text{area of sector } ADB .$$

Leibniz argued that

$$\text{area of sector } ADB = \text{area of } \triangle ADB + \frac{1}{2}(\text{area of region } AEL)$$

by his transmutation theorem (6.10). Explain how we may therefore conclude that

$$\arctan z = \frac{1}{2}y + \frac{1}{2}xz - \frac{1}{2}(\text{area of region } AHL) .$$

(c) After applying $y = z(2 - x)$, Leibniz concluded that

$$\arctan z = z - \frac{z^3}{3} + \frac{z^5}{5} - \frac{z^7}{7} + \cdots$$

using virtually the same arguments detailed in section 6.4. In your own words, explain the geometry underlying his argument.

Historical note. The series for the arctangent of an angle was previously known to Gregory, about two years prior to Leibniz, and to Nilakantha, about 150 years prior to Gregory. The argument of Leibniz presented here has the benefit of reminding us of his transmutation theorem.

9

Zero over zero

This chapter highlights the story of how scholars engaged in a conversation with the aim to remove from calculus all ambiguity, especially with regard to the infinitely large and infinitely small. Despite the common goal, there was at first no agreement on the solution, and the conversation meandered as most great debates do. Some ideas, dropped for dozens of years, made surprising returns in pamphlets, private letters, books, and book reviews. Ultimately, the voices in the debate unified on a course of action that banished ambiguity while opening marvelous new possibilities.

9.1 *D'Alembert and the convergence of series*

Jean le Rond d'Alembert (France, born 1717) played a key role in the eventual success of the drive toward rigor. To understand one of his contributions, recall Jyesthadeva's identity

$$\frac{1}{1+x} = 1 - x + x^2 - x^3 + x^4 - \cdots, \tag{9.1}$$

which we studied with the aid of Figure 2.5. When $0 < x < 1$, we are in the situation illustrated by the figure, so we can rest assured that if we substitute such a value for x on the right-hand side of (9.1), then the series will equal the left-hand side. This is what d'Alembert meant when he stated that the series in (9.1) *converges* to $1/(1+x)$ when $0 < x < 1$.

The sum in (9.1) clearly fails to converge to $1/(1+x)$ when $x \geq 1$, but the sum might converge for values of x less than zero. For example, we arrive at the true statement

$$2 = 1 + \frac{1}{2} + \frac{1}{4} + \frac{1}{8} + \cdots$$

123

by substituting $x = -1/2$ in (9.1). For $x = -2/3$, the statement

$$3 = 1 + \frac{2}{3} + \frac{4}{9} + \frac{8}{27} + \cdots$$

is plausible, but not immediately obvious.

Like always, it is best to stay cautious when making claims of truth about the infinite. In this case, we are aided by the fact that the right-hand side of (9.1) is a geometric series. Let P_n denote the partial sum obtained by interrupting the series at the x^n term, so

$$P_n = 1 - x + x^2 - x^3 + x^4 - \cdots \pm x^n \, ,$$

where the last sign depends on n being odd or even. Multiplying (and then dividing) through by $(1 + x)$ gives

$$P_n = \frac{1 \pm x^{n+1}}{1 + x} \, .$$

When $-1 < x < 1$, alternatively stated $|x| < 1$, the term x^{n+1} approaches zero as n approaches infinity, so the partial sums converge to $1/(1 + x)$. When $|x| \geq 1$, on the other hand, the partial sums diverge.

The series d'Alembert tackled posed bigger challenges, as they are not geometric. He investigated the convergence of identities like

$$(1 + x)^{1/2} = 1 + \frac{1}{2}x - \frac{1}{8}x^2 + \frac{1}{16}x^3 - \frac{5}{128}x^4 + \frac{7}{256}x^5 - \cdots \qquad (9.2)$$

that are generated by Newton's generalized binomial theorem. Recall that (3.16) counts the number of ways we may choose k out of n distinct objects, but also remember that we can use the more general (3.15) to calculate such oddities as

$$\binom{1/2}{4} = \frac{(1/2)(-1/2)(-3/2)(-5/2)}{4!}$$

$$= -\frac{5}{128}$$

to find the coefficient of x^4 in (9.2).

D'Alembert considered the ratio of consecutive terms of (9.2), comparing the typical term containing x^{n+1} to the term containing x^n that precedes it:

$$\frac{\left| \binom{1/2}{n+1} x^{n+1} \right|}{\left| \binom{1/2}{n} x^n \right|}$$

$$= \left| \frac{(1/2)(-1/2)\cdots(1/2-(n-1))(1/2-n)}{(1/2)(-1/2)\cdots(1/2-(n-1))} \cdot \frac{n!}{(n+1)!} \cdot x \right|$$

$$= \left| \frac{1/2-n}{n+1} \right| \cdot |x|$$

$$= \left| \frac{1/(2n)-1}{1+1/n} \right| \cdot |x| . \tag{9.3}$$

Interested in what ultimately happens to the ratio, d'Alembert observed that when n becomes arbitrarily large, the final expression in (9.3) simply equals $|x|$. He concluded that when $|x| \geq 1$, the ratio of each term in the series to the one before it is, ultimately, not diminishing, so there is no possibility that the right-hand side of (9.2) is finite. On the other hand, when $|x| < 1$, there is hope. Graphing $(1+x)^{1/2}$ along with a few partial sums of the series persuasively illustrates this conclusion.

Arguing more generally, d'Alembert showed that we must have $|x| < 1$ to have the chance of convergence for any expansion of the type $(1+x)^m$. The identity

$$(1+x)^m = \sum_{n=0}^{m} \binom{m}{n} x^n$$

from exercise 3.4(c) lets us compare consecutive terms as we did in (9.3):

$$\frac{\left| \binom{m}{n+1} x^{n+1} \right|}{\left| \binom{m}{n} x^n \right|}$$

$$= \left| \frac{(m)(m-1)(m-2)\cdots(m-n)}{(m)(m-1)(m-2)\cdots(m-(n-1))} \cdot \frac{n!}{(n+1)!} \cdot x \right|$$

$$= \left| \frac{m-n}{n+1} \right| \cdot |x|$$

$$= \left| \frac{m/n-1}{1+1/n} \right| \cdot |x| .$$

Letting n become arbitrarily large, as before, reveals that only if $|x| < 1$ is there hope that a series of the type $(1+x)^m$ can converge.

9.2   Lagrange defines the 'derived function'

The fairly narrow interval $|x| < 1$ suggests that an identity like (9.2) is not often useful. Might a different series approximate $(1 + x)^{1/2}$ for other values of x?

The discovery of such a series helped put series in the limelight; some scholars, in fact, believed that series could replace the "infinitely small" and "infinitely large" conundrums that were resisting all attempts to confirm them as logically possible. One such thinker, **Joseph Louis Lagrange** (France/Italy, born 1736) believed that *all* expressions involving one variable could be represented as *power series* — that is, infinite polynomials like the right-hand side of (9.2). In a way, he was correct, and we will return to this part of the tale in the next section.

For now, we return to the question posed in the first paragraph of this section, but phrase it with some new notation. When Lagrange became interested in these matters, he joined an ongoing conversation about *functions* that had begun in earnest in the writings of Leibniz. A function, for Leibniz, described a geometric object that inherits its features from another. The circle of Descartes in Figure 3.6, for example, inherits its position and radius based on the point we choose on the curve, so Leibniz would say that the circle is a function of the curve.

Later, in Leibniz's correspondence with Johann Bernoulli, the two applied this principle to variables: if one variable inherits its value thanks to some expression that involves another variable, then the former is a function of the latter. Focus shifted from the study of similar triangles to the study of curves, as evidenced by the nature of the figures in the latter chapters of this book.

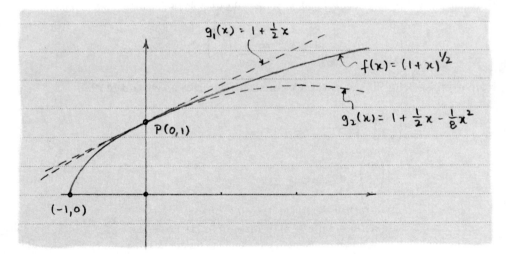

Figure 9.1. The functions g_1 and g_2 approximate f at the point P by matching successive derivatives.

When Lagrange wrote $f(x) = (1 + x)^{1/2}$, then, he meant that the function f takes on a value for each x that is given by the expression $(1 + x)^{1/2}$; for example, $f(3) = (1 + 3)^{1/2} = 2$. Sometimes we cannot evaluate a function at all x values, as with f when $x < -1$. This may not strike a modern reader as a great leap, but functions proved a fertile symbolic invention, and the care with which mathematicians pinned down the meaning of *function* opened up many other conversations that helped remove illogical notions from calculus.

We may use this notation to discover (9.2) using a new approach. Suppose we wish to approximate $f(x) = (1 + x)^{1/2}$ near the point $P(0, 1)$ shown in Figure 9.1, and we wish our approximating function to be

$$g(x) = c_0 + c_1 x + c_2 x^2 + c_3 x^3 + \cdots ,$$

where the coefficients $c_0, c_1, c_2, c_3, \ldots$ are unknown. (Thanks of course to (9.2), we do know these coefficients, but we will start from scratch.) We can build g term by term by considering its partial sums. Using subscripts to identify the stopping points in these partial sums, we can write

$$g_1(x) = c_0 + c_1 x$$

and ask what c_0 and c_1 ought to equal so that g_1 best approximates f at P. Requiring g_1 to pass through P seems quite wise, so we solve

$$g_1(0) = f(0) \Longrightarrow c_0 = 1, \text{ so } g_1(x) = 1 + c_1 x .$$

The coefficient c_1 controls the slope of the line that g_1 describes. If we choose c_1 so that this line is tangent to f at P, we arguably get the line that best approximates f at that point. Not only is this merely a claim, but also it is vague, as we might disagree on what "best" means.

Still, it is a reasonable claim, so we pursue it. In doing work similar to this, Lagrange argued that c_1 can be calculated with a function that is *derived* from f and is denoted f' to indicate this. From this verb, we get the term *derivative*. This derivative f' is equivalent to Leibniz's ratio dy/dx where $y = (1 + x)^{1/2}$, because this ratio also gives the slope of tangent lines. Thus, we know that

$$f(x) = (1 + x)^{1/2} \Longrightarrow f'(x) = \frac{1}{2}(1 + x)^{-1/2} ,$$

using the chain rule (see exercise 7.6).

Because $g_1'(x) = c_1$ and we want $g_1'(0) = f'(0)$, we conclude that $c_1 = 1/2$. So $g_1(x) = 1 + (1/2)x$ stands as our best approximation to f at P thus far. Compare the partial sum g_1 with (9.2).

The next partial sum $g_2(x) = c_0 + c_1 x + c_2 x^2$ has an extra term, so we hope it will improve upon our first approximation. Repeating the argument we just studied for creating g_1 results in the same conclusions $c_0 = 1$ and $c_1 = 1/2$, so $g_2(x) = 1 + (1/2)x + c_2 x^2$. The functions g_2 and f not only agree at P but also share the same rate of change at that point, as given by their derivatives.

Observe how Figure 9.1 shows that the rate of change of g_1 is constant at P, while the rate of change of f is not constant, but is decreasing. Perhaps we can use the extra coefficient c_2 in g_2 to adjust to this fact.

The rate of change of f is given by its derivative f', so to learn about the rate of change of f' we calculate its derivative

$$f'' = -\frac{1}{4}(1 + x)^{-3/2}.$$

Setting the 'second' derivatives of f and g_2 equal at P gives

$$g_2''(0) = f''(0) \Longrightarrow 2c_2 = -\frac{1}{4} \Longrightarrow c_2 = -\frac{1}{8}.$$

Again we compare our results favorably to (9.2). Figure 9.1 illustrates how g_2 tends to more closely approximate f near P than did g_1. Following this approach to find the partial sums $g_3, g_4, \ldots$, we rediscover (9.2).

9.3 Taylor approximates functions

Now that we are warmed up, we step back a bit in time to study an important result found by **Brook Taylor** (England, born 1685). Having jumped past Taylor to arm ourselves with function notation, we have streamlined our task.

We will examine our specific question from the opening paragraph of section 9.2 before looking at Taylor's general result. How might we approximate $f(x) = (1 + x)^{1/2}$ with a power series at a point other than P? We choose the point $Q(3, 2)$ for

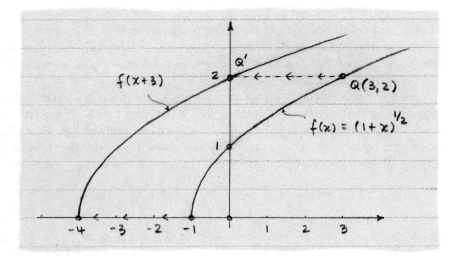

Figure 9.2. To approximate a function with a power series at a point where $x \neq 0$, we translate the function first.

convenience, shown in Figure 9.2. We could make use of our approach in section 9.2 if we slide f horizontally until Q lies on the vertical axis at Q'. The function $f(x + 3)$ accomplishes this *translation* of f. We may approximate the translated function at Q' as

$$f(x + 3) = c_0 + c_1 x + c_2 x^2 + c_3 x^3 + \cdots,$$

and then slide our approximation and function back to its original position. The re-translated approximation

$$\begin{aligned} f(x) &= f((x - 3) + 3) \\ &= c_0 + c_1(x - 3) + c_2(x - 3)^2 + c_3(x - 3)^3 + \cdots \end{aligned} \tag{9.4}$$

will then be aimed at point Q. Letting g denote the series in (9.4), we find that

$$f(3) = g(3) \implies c_0 = 2 \, ,$$
$$f'(3) = g'(3) \implies c_1 = 1/4 \, ,$$
$$f''(3) = g''(3) \implies c_2 = -1/64 \, ,$$

and so on, leading to the approximation

$$g(x) = 2 + \frac{1}{4}(x - 3) - \frac{1}{64}(x - 3)^2 + \frac{1}{512}(x - 3)^3 - \cdots \, . \qquad (9.5)$$

If we expand each power of $(x - 3)$ and collect like terms, we have the desired power series. To avoid this (infinite) labor, mathematicians agreed to call a series like (9.5) a *power series about* $x = 3$.

Taylor's general argument was more along the lines of Newton than Lagrange; in fact, Newton was not alone in having knowledge of power series before Taylor published his result. However, the same sort of reasoning we have used in this section produces the conclusion that these mathematicians reached: if we wish to approximate a function f with a power series about a by finding the unknown coefficients in

$$f(x) = c_0 + c_1(x - a) + c_2(x - a)^2 + c_3(x - a)^3 + c_4(x - a)^4 + \cdots \, ,$$

we may evaluate the derivatives of each side of this equality at $x = a$ and equate them. The first few steps are

$$f'(x) = c_1 + 2c_2(x - a) + 3c_3(x - a)^2 + 4c_4(x - a)^3 + \cdots \, , \text{ so}$$
$$f'(a) = c_1 \, ,$$

$$f''(x) = 2c_2 + (3 \cdot 2)c_3(x - a) + (4 \cdot 3)c_4(x - a)^2 + \cdots \, , \text{ so}$$
$$f''(a) = 2c_2 \, ,$$

$$f'''(x) = (3 \cdot 2)c_3 + (4 \cdot 3 \cdot 2)c_4(x - a) + \cdots \, , \text{ so}$$
$$f'''(a) = (3 \cdot 2)c_3 \, .$$

Writing $f^{(n)}$ for the nth derivative of f to keep the notation tidy, we conclude that

$$f^{(n)}(a) = n! \cdot c_n$$

in the general case. So if f has infinitely many derivatives at the point where $x = a$, we conclude with Taylor that the *power series generated by* f *at* $x = a$ is

$$f(x) = \sum_{k=0}^{\infty} \frac{f^{(k)}(a)}{k!}(x - a)^k \, . \qquad (9.6)$$

One benefit of power series is the ease with which we can differentiate and integrate them. In Chapter 5, for example, we studied the link between logarithms and the quadrature of the hyperbola $1/(1 + x)$. Then, in Chapter 7, we connected quadrature and integration. These matters come together clearly thanks to the series

$$\log(1 + x) = x - \frac{x^2}{2} + \frac{x^3}{3} - \frac{x^4}{4} + \cdots \tag{9.7}$$

and

$$\frac{1}{1 + x} = 1 - x + x^2 - x^3 + x^4 - \cdots , \tag{9.8}$$

which give us strong reason to believe that

$$\int \frac{1}{1 + x} \, dx = \log(1 + x) .$$

The trickiest issues with power series involve convergence. As we have seen, some series give a terrific approximation but only within a narrow interval; outside the interval $|x| < 1$, for example, the series on the right-hand side of (9.8) utterly fails to approximate $1/(1 + x)$. If we integrate (9.8), then, to get (9.7), does the new series converge on the same interval? And if we want to *use* (9.7) (not simply *admire* it) to get a good approximation of $\log(1 + x)$, then what is the shortest partial sum of the series that will get us close enough?

Almost any door that opens in mathematics leads to others, some locked more tightly than the first. This door resisted mathematical lockpicking for decades: what *is* convergence?

9.4   Bolzano and Cauchy define convergence

Many thinkers, including d'Alembert and Lagrange, tried their hand at defining convergence, often in reaction to failed attempts on the part of their predecessors. One overarching theme of the debate centered on criticisms leveled against calculus as presented by its initial explorers, Newton and Leibniz. Both men struggled to rid their arguments of the ambiguity and illogic introduced by the infinitely large, infinitely small, and (worse yet) ratios of such quantities. The ratio of differentials dy/dx was particularly puzzling because of its usefulness in answering all manner of real-world problems; the trouble, in a nutshell, is that to *be* useful, the ratio had to shift into a mysterious world where it equaled zero divided by zero.

In one attempt to defuse this issue, Leibniz argued that the ratio dy/dx gives the slope of a tangent line to the curve y, and because this slope is often a non-mysterious number, then clearly dy/dx must not equal $0/0$. Newton defended his calculus in similar ways. Tangent lines illustrate the issue well; as in Figure fg-DescartesTangentLine, some mathematicians considered tangent lines to be the ultimate fate of a sequence of secant lines through points that are approaching each other ever more closely. The words "ultimate" and "approaching" and "closely" seemed key.

These concepts are also at the forefront when we consider the convergence of series like (8.13), reproduced here:

$$\sin \alpha = \alpha - \frac{\alpha^3}{3!} + \frac{\alpha^5}{5!} - \frac{\alpha^7}{7!} + \cdots . \tag{9.9}$$

If we claim that the series in (9.9) converges to $\sin \alpha$ for a particular angle, say, $\alpha = \pi/2$, then what do we mean? It is impossible to substitute $\pi/2$ for α infinitely many times, so we must rely on partial sums of the series. To this end, we let g_k be the partial sum that has k non-zero terms, and evaluate the first few:

$$g_1(\alpha) = \alpha \implies g_1(\pi/2) \approx 1.57079632,$$

$$g_2(\alpha) = \alpha - \frac{\alpha^3}{3!} \implies g_2(\pi/2) \approx 0.92483222,$$

$$g_3(\alpha) = \alpha - \frac{\alpha^3}{3!} + \frac{\alpha^5}{5!} \implies g_3(\pi/2) \approx 1.00452485,$$

$$g_4(\alpha) = \alpha - \frac{\alpha^3}{3!} + \frac{\alpha^5}{5!} - \frac{\alpha^7}{7!} \implies g_4(\pi/2) \approx 0.99984310,$$

$$g_5(\alpha) = \alpha - \frac{\alpha^3}{3!} + \frac{\alpha^5}{5!} - \frac{\alpha^7}{7!} + \frac{\alpha^9}{9!} \implies g_5(\pi/2) \approx 1.00000354.$$

Thus far, the partial sums appear to give better and better approximations of $\sin(\pi/2) = 1$; that is, the sequence

$$g_1(\pi/2), \; g_2(\pi/2), \; g_3(\pi/2), \; g_4(\pi/2), \; g_5(\pi/2), \; \ldots \tag{9.10}$$

seems to converge to 1.

The first scholar to provide a clear definition of convergence was **Bernard Bolzano** (Bohemia, born 1781), who was a toddler when d'Alembert died in his sixties. Why a solid definition of convergence took so many years to craft is due in part to the potential pitfalls; for example, how do we know that the partial sums beyond those we have checked continue to approach 1? What if, in some unexplored part of the sequence (9.10), the partial sums unexpectedly deviate from 1, perhaps to go on deviating, or perhaps ultimately to approach 1 after all?

Bolzano deftly navigated these troubles with his criterion for convergence:

> For a sequence like (9.10) to converge, there must come a place in the sequence beyond which every pair of values in the sequence differ by as little as we choose. $\tag{9.11}$

The phrase "as little as we choose" acts like a noose, tightening on the infinitely long tail end of the sequence.

The noose is a positive number, usually denoted with the Greek letter ϵ (read 'epsilon') in reference to the word 'error'. All Bolzano required is that $\epsilon > 0$, so we may draw the noose as tightly as we like. Whenever we stop at a particular partial sum to evaluate it at $\alpha = \pi/2$, its value may err from 1. But if there is always a point in the sequence beyond which all of the values are within ϵ of each other, then the sequence converges. Further, if there is always a point in the sequence beyond which all of the values are within ϵ of some constant, then the sequence converges to that constant.

This last statement is due to **Augustin Cauchy** (France, born 1789), whose careful definitions and proofs are the models for how calculus is often taught today.

His treatment of calculus pushed the subject from geometry to algebra. Consider this algebraic expression of Cauchy's definition of the convergence of a sequence:

> A sequence $u_1, u_2, u_3, \ldots$ converges to L if and only if for every $\epsilon > 0$ there is some m such that
>
> $$|u_k - L| < \epsilon \text{ for all } k \geq m .$$

(9.12)

For thinkers well-versed in geometry, this statement may seem forebodingly abstract; but at least we can check it. This definition mathematically captures the essence of words like "ultimately" and "approaches" while avoiding any reliance on drawings.

Borrowing a term already in use, Cauchy called L the *limit* of the sequence; for example, we suspect that the sequence (9.10) is converging to the limit $L = 1$. To confirm our suspicion, we would need to show that no matter how tiny we choose $\epsilon > 0$, we can identify the 'tail end' of the sequence in which every term is within ϵ of the limit 1. Although this task may not always be simple, it is a clearly-stated mathematical objective that is in harmony with our intuitions about convergence.

We may also define when a series like (9.9) converges using the precise language of algebra:

> Let P_j denote the jth partial sum of the series $\sum_{k=1}^{\infty} a_k$. Then the series converges if and only if for every $\epsilon > 0$ there is some i such that
>
> $$|P_n - P_m| < \epsilon \text{ for all } n, m \geq i .$$

(9.13)

Reminiscent of Bolzano's (9.11), this definition states that the convergence of a series depends on its partial sums becoming arbitrarily close.

For example, thanks to the figure in exercise 2.3, we suspect that when $|x| < 1$ we have

$$\frac{1}{1-x} = 1 + x + x^2 + x^3 + \cdots + x^n + \cdots .$$

Our hunch is confirmed by (9.13); letting $n > m$,

$$
\begin{aligned}
|P_n - P_m| &= |x^{m+1} + x^{m+2} + x^{m+3} + \cdots + x^n| \\
&= |x^{m+1}(1 + x + x^2 + \cdots + x^{n-(m+1)})| \\
&= \left| x^{m+1} \cdot \frac{1 - x^{n-m}}{1-x} \right|, \text{ as on page 124} \\
&= \frac{1}{1-x} |x^{m+1} - x^{n+1}| \\
&\leq \frac{1}{1-x} |x^n - x^{n+1}| \\
&= |x^n|,
\end{aligned}
$$

and by choosing n large enough we can force $|x^n| < \epsilon$ for any $\epsilon > 0$.

Although Cauchy did not express (9.13) in quite its technical way, this check on the convergence of series is often called "the Cauchy convergence criterion" (see exercise 9.5). Note that it merely stipulates when a series does converge, but not to what limit. Finding the limit itself is a further puzzle.

By such means, Bolzano and Cauchy began what amounted to a housecleaning of calculus, brushing away the clutter down to the hardwood floor and then starting over. They reused plenty of the existing notation and arguments, but they and their peers sharpened concepts, tightened logic, and published books. As is so often the case, these efforts spurred others to explore whatever weaknesses might linger. Mathematical exploration often takes the form of surprising examples; we will take a look at a few of these in our final chapter.

9.5 *Furthermore*

9.1 **D'Alembert illuminates the question of convergence.** To motivate his results on the convergence of (9.2), d'Alembert experimented with an x value just a bit larger than 1; he chose $x = 200/199$, but we will start with the more modest value $x = 4/3$. Because this value is larger than 1, we know that (9.2) is not valid. Nevertheless, because $4/3$ is not *much* larger than 1, the first few partial sums (9.2) approximate $(1 + 4/3)^{1/2}$ fairly well.

   (a) D'Alembert let n become arbitrarily large in (9.3) to investigate what values of x would cause the series in (9.2) to diverge. Flipping this approach on its head, we have chosen $x = 4/3$ and now ask for what values of n will the final expression in (9.3) exceed 1. What is your answer?

   (b) Recall that n is the exponent of a typical term in the series in (9.2). Using $x = 4/3$ and your value of n from part 9.1(a), calculate the first $n + 2$ partial sums of the series. What do you notice about the partial sums?

   (c) Answer part 9.1(a) for d'Alembert's value $x = 200/199$. His point in choosing a number just a tiny bit larger than 1 was to illustrate the danger of believing what you see (in mathematics) without proving your belief. The partial sums in (9.2) grow quite close to $(1 + 200/199)^{1/2}$ for quite some time before eventually failing to converge. Which phrase – "quite close to" or "quite some time" – does your solution help quantify?

9.2 **Approximating functions.** Continue to approximate $(1 + x)^{1/2}$ as in section 9.2 by finding g_3 and g_4.

9.3 **Taylor's method justifies Jyesthadeva's conclusion.** Use (9.6) to rediscover (9.8), which Jyesthadeva found by other means in chapter 2.

9.4 **A Taylor polynomial for $\sin \alpha$.** In chapter 8 we saw how Jyesthadeva discovered an infinite polynomial (8.13) that is equivalent to the function for the sine of an angle. Using (9.6), draw the same conclusion.

9.5 **The Cauchy convergence criterion.** In his book *Cours d'analyse,* Cauchy defined the convergence of series as (9.13) does. He let

$$s_n = u_1 + u_2 + u_3 + \cdots + u_{n-1}$$

denote the partial sum of an infinite series, then highlighted "the successive differences between the first sum s_n and each of the following sums" as determined by the equations

$$s_{n+1} - s_n = u_n,$$
$$s_{n+2} - s_n = u_n + u_{n+1}, \qquad\qquad (9.14)$$
$$s_{n+3} - s_n = u_n + u_{n+1} + u_{n+2},$$

and so on. Cauchy stated that the infinite series converges so long as the sums on the right-hand side of (9.14) "eventually constantly assume numerical values less than any assignable limit." The more symbolic (9.13) removes any ambiguity lurking in these words.

(a) **A geometric series.** We immediately used (9.13) on page 132 to discuss the convergence of the geometric series

$$\frac{1}{1-x} = 1 + x + x^2 + x^3 + \cdots + x^n + \cdots,$$

where $|x| < 1$. Cauchy's treatment uses (9.14); he pointed out that the finite sums starting with x^n may be algebraically rewritten

$$x^n + x^{n+1} = x^n \, \frac{1 - x^2}{1 - x},$$
$$x^n + x^{n+1} + x^{n+2} = x^n \, \frac{1 - x^3}{1 - x},$$

and so on. He claimed that each of these sums is therefore contained between

$$x^n \quad \text{and} \quad \frac{x^n}{1 - x}.$$

Why may he claim this, and how does this observation connect to his definition of convergence to prove that the series converges?

(b) **The series for e.** As another application of his definition, Cauchy examined the series

$$1 + \frac{1}{1!} + \frac{1}{2!} + \frac{1}{3!} + \cdots + \frac{1}{n!} + \cdots,$$

previously seen in (6.23). As in part (a), Cauchy rewrote the finite sums

$$\frac{1}{n!} + \frac{1}{(n+1)!},$$
$$\frac{1}{n!} + \frac{1}{(n+1)!} + \frac{1}{(n+2)!},$$

(and so on) in such a way that convinced him that any of these sums "decreases indefinitely as n increases." He uses what we now call the Comparison Test, which is to say that he finds a convergent series that term-by-term is greater than any sum in the list above. As a hint, notice that

$$\frac{1}{(n+1)!} < \frac{1}{n!} \cdot \frac{1}{n},$$

and try to recapture his argument yourself.

10
Rigor

Despite its deductive nature, mathematics yields its truths much like any other intellectual pursuit: someone asks a question or poses a challenge, others react or propose solutions, and gradually the edges of the debate are framed and a vocabulary is built. One might attempt to distinguish mathematics from other disciplines by arguing that, ultimately, we *know* that its results express *truth* in a way no other subject can boast; however, philosophical arguments of the early 1900s call even this claim into question.[1]

While the story of calculus features plenty of intrigue and debate, readers should rest assured that controversy is often a hallmark of mathematical discovery. As with every pursuit of the mind, mathematics advances when its explorers resist attempts to settle matters. Think of cubism challenging the dominance of linear perspective in Western art, jazz releasing music from traditional notions about rhythm and harmony, or free verse perturbing the boundaries of poetic meter. Every intellectual pursuit needs people who respect the rules yet stretch them.

Calculus shifted in its focus from geometry and puzzling claims about the infinite to functions, limits, and picture-free algebra. During the transition, each proposed definition or rule weathered a barrage of exceptions. The goal of the debate was certainty.

10.1 Cauchy defines continuity

For thousands of years, from Archimedes to Newton, anyone discovering a curve likely did so via geometry or observations of nature. The Greeks, for example, considered a parabola to be the intersection of a plane with a cone, as if they felt a pull to study only those curves that result when simpler objects interact. The

---

[1]Rebecca Goldstein's *Incompleteness: The proof and paradox of Kurt Gödel* (New York: W. W. Norton, 2005) relates Gödel's striking demonstration that proof and truth are not synonymous even in mathematics.

cycloid studied by Galileo and Roberval is a by-product of a rolling circle. An acquaintance of Leibniz challenged him to discover the curve traveled by a watch as it is dragged by its chain across a table in a particular way. For years in the 1700s, scholars investigated the curves attained by an elastic that is stretched taut and then allowed to relax.

Euler exhorted his contemporaries to free themselves from the urge to base mathematics on the physical or geometric. The tool he championed was the *function*. A function acts like a set of rules for turning some numbers into others, a machine with parts that we can manipulate to accomplish anything we can imagine.

To drive this point home, we can invent a function and study it. Let $g(x)$ equal x when x is in the list $1, 1/2, 1/4, 1/8, \ldots$ and equal zero otherwise. A modern way to define this function is

$$g(x) = \begin{cases} x & \text{when } x = 1/2^m \text{ for } m = 0, 1, 2, \ldots, \\ 0 & \text{otherwise.} \end{cases}$$

Figure 10.1 is a good attempt at drawing it. Although we cannot depict the behavior of g near the origin, our imaginations can take over.

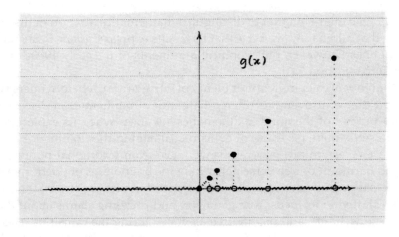

Figure 10.1. It is not possible to draw the points of g that lie near the origin.

At its non-zero points, the function 'jumps' from the horizontal axis. The open circles along the horizontal emphasize this. Our intuition suggests that a function is 'continuous' when it avoids jumps, and not continuous when it jumps. Many pieces of the function, like those between the open circles, appear continuous; does this hold as we imagine the function's behavior near the origin in Figure 10.1? The term 'continuous' begs for definition, lest we lose ourselves in fruitless speculation.

If we wish the word 'continuous' to prohibit jumps in a function, its definition must somehow control the vertical change of the function at a sort of microscopic level. That is, at any point on a 'continuous' function, the nearby points ought to be as 'close' as possible. When inspecting a particular point, we would be wise

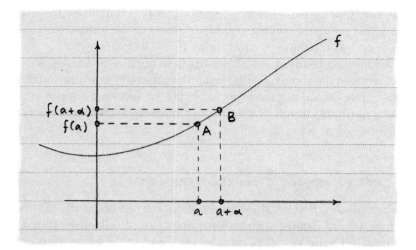

Figure 10.2. For f to be continuous where $x = a$, the vertical distance between A and B must vanish as the horizontal distance vanishes.

to heed the lesson of Bolzano and Cauchy at the end of section 9.4: do not look directly at the point, but look at how nearby points behave.

Here is Cauchy's version. To check if the function f in Figure 10.2 is continuous at the point where $x = a$ (corresponding to the point A), assign an increment α to a and locate the corresponding point B. As α vanishes, watch the behavior of B to see if its vertical difference from A also vanishes. If this difference $f(a + \alpha) - f(a)$ does not vanish, then f has a jump at A and is thus not continuous at $x = a$. The same should be true if we let α be negative.

Rather than use "vanish", Cauchy used the phrase "decreases indefinitely"; both refer to the limit concept we looked at in section 9.4. As then, we want an algebraic way to draw a noose around the variable a and its corresponding value $f(a)$. To put "the difference $f(a + \alpha) - f(a)$ vanishes as α vanishes" in algebraic terms, we may write

$$\lim_{\alpha \to 0} \big[f(a + \alpha) - f(a) \big] = 0 . \tag{10.1}$$

Cauchy condensed these observations into a definition of the continuity of a function: Given a function f, choose any values a and α. The function $f(x)$ is *continuous* between $a - \alpha$ and $a + \alpha$ if (10.1) is true. Using his definition to investigate g depicted in Figure 10.1, we conclude that g is not continuous at any of the values in the list $1, 1/2, 1/4, 1/8, \ldots$, but that g is continuous at all other non-zero values, as our intuition about g would lead us to believe.

What about where $x = 0$? To the left of the origin, all is calm, while to the right, matters are infinitely more interesting. If we look along any horizontal distance α to the right of the origin, we find infinitely many values of x where g is not continuous. Tracing along g from left to right, it seems as if the origin is where the myriad discontinuities begin.

Tracing toward the origin from the right, however, we see that the heights of the jumps are decreasing indefinitely. Every sequence $x_1, x_2, x_3, \ldots$ of x values that approaches zero gives a sequence $g(x_1), g(x_2), g(x_3), \ldots$ that also approaches zero, even if the latter sequence contains infinitely many of the jump points. Odd as it may seem, we conclude that g is continuous at $x = 0$.

One trick to seeing beauty in mathematics is to nurture this sense of "odd as it may seem" while at the same time understanding the subject well enough to know that oddities arise *despite* our attempts to set the subject on a simple, straightforward footing. With this in mind, we turn to a marvelous creation of Bolzano that challenges intuition at every turn.

10.2 *Bolzano invents a peculiar function*

When mathematicians first became curious about the rate of change of curves, they associated the rate of change with tangent lines. Specifically, they found the rate of change of f in Figure 10.2 at point A by considering the behavior of the secant line through A and B as α vanishes. As we have seen (in sections 3.3 and 9.4, for example), this approach leads to mistreating α as being both zero and not zero, or to puzzling ratios that equal $0/0$.

Cauchy sidestepped these issues by defining the derivative of a function as a limit. Note that this idea does not radically challenge our intuition about derivatives; the innovation here lies in the role of limits as a foundation for other concepts. So when Cauchy defined the derivative of a continuous function f as

$$\lim_{\alpha \to 0} \frac{f(a + \alpha) - f(a)}{\alpha}, \tag{10.2}$$

at the point where $x = a$, he requires us only to find the limiting value of the slope of the secant line. As Cauchy put it,

> But though these two terms [the numerator and denominator in (10.2)] will approach the limit zero indefinitely and simultaneously, the ratio itself can converge towards another limit, be it positive or negative.

Newton and Leibniz advanced similar arguments, but they lacked an algebraic definition of limit.

A function may be continuous on an interval without being differentiable at every point in the interval. In Figure 10.3, we see an example of this at the point $(0, 0)$ for the function $h(x) = |x|$. Checking (10.1), we find that h is continuous on every interval containing $x = 0$. When we check (10.2), however, the answer is not as clear. If $\alpha > 0$, we are considering the slopes of secant lines to the right of the origin; all of these slopes are 1, so we conclude that the limit is 1. If $\alpha < 0$, our investigation shifts to the left of the origin, where the corresponding limit is -1. These two limits do not match, so we say that the function is not differentiable at $x = 0$.

Replicating the 'sharp' point in Figure 10.3, we can create a function that is nondifferentiable at an infinite number of points; Figure 10.4 shows one possibility. Merely imagining and drawing a function, however, does not guarantee its existence. We should define it mathematically; after all, the function extends outside

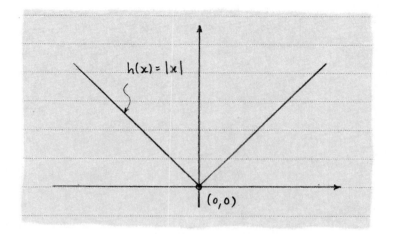

Figure 10.3. The function h is continuous at $(0, 0)$ but has no derivative there.

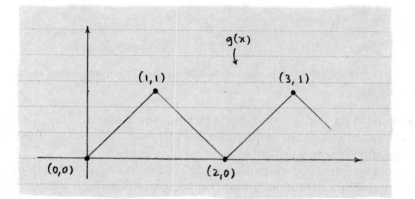

Figure 10.4. This function has infinitely many non-differentiable points.

the figure. We can define g with the help of the *ceiling function* $\lceil x \rceil$, which equals the smallest integer that is larger than x: for $x \geq 0$, let

$$g(x) = \begin{cases} \lceil x \rceil - x & \text{when } \lceil x \rceil \text{ is even,} \\ x - \lceil x - 1 \rceil & \text{when } \lceil x \rceil \text{ is odd.} \end{cases}$$

This function has infinitely many non-differentiable points while maintaining continuity everywhere.

In between the points where g is not differentiable are intervals where it is. Even if we completely unleash our imaginations, it is difficult to see how this could fail to be. How could two non-differentiable points crowd so closely together that there is no differentiable point between them, while the function is continuous at both points?

In an astonishing flight of fancy, Bolzano accomplished just this not only for two points, but for *all* points on his function. That is, he invented a function that is continuous everywhere, but so 'sharp' at every point that it is impossible to draw

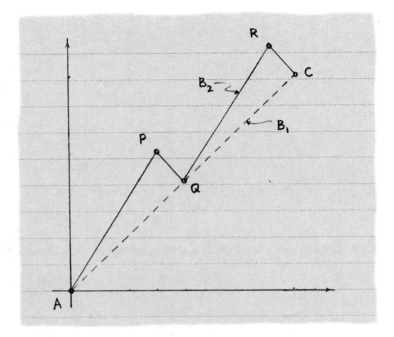

Figure 10.5. The shape of B_2 depends on that of B_1, and this contingency continues for functions $B_3, B_4, \ldots$.

a tangent line to the curve anywhere. In fact, it is impossible to draw his function at *all*; we can only draw the stepping stones that lead us there, and then rely on our imaginations to carry us further.

A few paragraphs ago, we paused to define the function g (shown in Figure 10.4), hinting that a definition is more rigorous and clear than a mere drawing. But Bolzano's function *cannot* be drawn; it can only exist by virtue of its definition. How appealing it is that a formal definition, *not* a drawing, is what freed Bolzano's imagination to construct his fantastical function.

As you might suspect, we cannot define this function in a single step. Bolzano created a variation of the function in Figure 10.4 by adding more 'sharp' points at each step of an infinite process, aiming for a function B that is itself a limit. We restrict ourselves here to defining B on x values between 0 and 1, although it is possible to extend his idea to any set of x values.

Begin with the simple function $B_1(x) = x$, depicted with a dotted line in Figure 10.5. The next function B_2 appears as a solid line and is constructed from B_1 according to the following recipe: Let $A(a,b)$ and $C(c,d)$ be the endpoints of any line segment that is part of the function B_k. (So A is $(0,0)$ and C is $(1,1)$ for B_1.) Between $x = a$ and $x = c$ we will select three values at which the next function B_{k+1} will be non-differentiable. The three points corresponding to these values are

$$P = \left(a + (3/8)(c - a), b + (5/8)(d - b)\right),$$
$$Q = \left(a + (1/2)(c - a), b + (1/2)(d - b)\right),$$
$$R = \left(a + (7/8)(c - a), d + (1/8)(d - b)\right).$$

(For B_2, these points are $P(3/8, 5/8)$, $Q(1/2, 1/2)$, and $R(7/8, 9/8)$.) The segments AP, PQ, QR, RC together become a part of the function B_{k+1}. This process is repeated for all segments of B_k, and then the process begins anew with B_{k+1}.

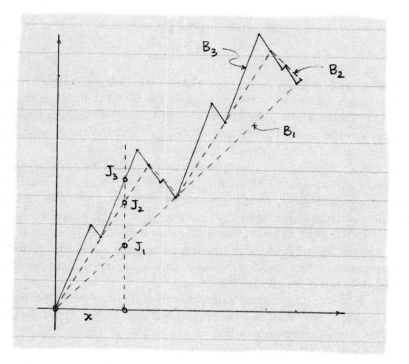

Figure 10.6. The sixteen pieces of B_3 (solid line) are created from the four pieces of B_2 (dashed).

Figure 10.6 shows the result when this process is used to modify the segments of B_2 to create B_3. Each step increases the number of "rises and falls," as Bolzano put it. His process aims to create a function B that is the limit of the functions $B_1, B_2, B_3, \ldots$ in the same way that the limit of the partial sums of the series in (8.13) or (9.8) can be the functions $\sin \alpha$ or $1/(1 + x)$. The function B has a remarkable property: no matter how closely we inspect it, we see only the sharp turning points that signal the rises and falls; yet we never see a single *example* of a rise or fall. It is as if the turning points have crowded out all of the segments connecting them. Nevertheless, *at the same time*, the function B is continuous; the turning points are somehow infinitely close to one another as B zigzags its way from $(0, 0)$ to $(1, 1)$.

The verb "zigzags" fails to do justice, as it makes us think that B can be drawn. But despite our inability to draw it, we can define B and argue about it, which was Bolzano's purpose in imagining it. He designed his function as a stress test on intuitive definitions of 'continuous.' A function such as B could break a person's faith in these intuitions.

Bolzano first proved that B exists; after all, his process merely defines the stepping stones that lead to it. He then shows that the function is continuous at every point. Finally, he explains why we cannot — no matter how closely we look — find even a single tiny piece of B that is rising or falling. We will look at these three arguments in turn.

Choose any x between 0 and 1 and consider the sequence

$$B_1(x), B_2(x), B_3(x), \dots$$

as depicted by the points $J_1, J_2, J_3, \dots$ in Figure 10.6. Bolzano claimed that every such sequence approaches a limit; that is, appealing to (9.11), Bolzano argued that there comes a place in the sequence beyond which all values are indefinitely close to each other. In other words, the points $J_1, J_2, J_3, \dots$ eventually crowd arbitrarily close to the point that Bolzano defines to be on the function B.

Because this approach only considers vertical distances, let us define the *height* of any piece of B_k to be the vertical difference of its endpoints. For example, two of the four pieces of B_2 have height 5/8 and the other two have height 1/8. This caps the vertical distance between any point on B_1 and the corresponding point on B_2 at 5/8. In fact, we can see in Figure 10.6 that if we slide the vertical line at x through the entire figure, then the maximum vertical distance between the points J_1 and J_2 is quite a bit less than 5/8. However, all Bolzano needs is a cap that does the job, so it does not matter that 5/8 is generous.

Similarly, when we construct the sixteen pieces of B_3 from the four pieces of B_2, the largest height of any piece of B_3 is (much) less than $(5/8)^2$. This caps the maximum vertical distance between points like J_2 and J_3 in Figure 10.6. In general, this cap will equal $(5/8)^n$ as we construct B_{n+1} from B_n.

So if we think about the general point J_n on the function B_n, and extend our thinking r steps further where we would reach the point J_{n+r} on B_{n+r}, we can conclude that the vertical difference between J_n and J_{n+r} is capped by the sum of the caps at each of the r intermediate steps; this sum is

$$\left(\frac{5}{8}\right)^n + \left(\frac{5}{8}\right)^{n+1} + \left(\frac{5}{8}\right)^{n+2} + \cdots + \left(\frac{5}{8}\right)^{n+r-1} . \tag{10.3}$$

Now (10.3) is certainly smaller than the infinite sum

$$\left(\frac{5}{8}\right)^n + \left(\frac{5}{8}\right)^{n+1} + \left(\frac{5}{8}\right)^{n+2} + \cdots + \left(\frac{5}{8}\right)^{n+r-1} + \left(\frac{5}{8}\right)^{n+r} + \cdots ,$$

which simplifies to

$$\frac{8}{3}\left(\frac{5}{8}\right)^n \tag{10.4}$$

with the help of (9.1), where $x = -5/8$. Because (10.4) decreases indefinitely with the indefinite increase of n, we can follow Bolzano in claiming that the points $J_1, J_2, J_3, \dots$ approach a point J that has a particular value, no matter what value of x we start with. Bolzano's function B, being composed of all such points J, therefore exists.

Bolzano dispatched with the argument that B is continuous by pointing out that B is the limit of the functions $B_1, B_2, B_3, \ldots$ as just described, and these functions are all obviously continuous themselves. Put a mental bookmark here; we will return.

Last, Bolzano examined the turning points of B by considering the horizontal features of his sequence of functions. In contrast with our definition of the *height* of a piece of B_k, we will say that the horizontal distance between the endpoints of a piece of B_k is its *width*. Of the four pieces of B_2, two have width 3/8 and two have width 1/8. The maximum width of any of the sixteen pieces of B_3 is $(3/8)^2$, and so on. Because $(3/8)^n$ vanishes with the indefinite increase of n, we know that the maximum width of any piece will decrease indefinitely. That is, for any $\epsilon > 0$, we can find an n large enough so that $(3/8)^n < \epsilon$.

Thus, if we look at any x between 0 and 1 and consider values between $x - \epsilon$ and $x + \epsilon$, some function in the sequence $B_1, B_2, B_3, \ldots$ will exhibit turning points at some of these values (as will all subsequent functions in the sequence). This guarantees the marvelous property of B that, in a sense, it consists *only* of turning points.

What else can one do but take a deep breath and wonder at how mathematics seems to offer up its most profound and bizarre truths precisely when its practitioners are trying to pare it down to its most elemental, intuitive building blocks?

10.3   Weierstrass investigates the convergence of functions

Bolzano's function B can boast all of the characteristics he claimed of it, but his argument in support of continuity was too hasty. Recall that he believed that B is continuous simply because it was the limit of functions $B_1, B_2, B_3, \ldots$ that are all continuous. This completely plausible belief is in error.

We can demonstrate this by finding a sequence of continuous functions that does *not* have a continuous limit. The sequence

$$\cos \alpha, (\cos \alpha)^2, (\cos \alpha)^3, (\cos \alpha)^4, \ldots \qquad (10.5)$$

does the trick; each of these functions is continuous, yet the limit of the sequence is not. Consider the functions only on the interval between $A(-\pi/2, 0)$ to $B(\pi/2, 0)$ as in Figure 10.7; at all values except $\alpha = 0$, the function $\cos \alpha$ takes on non-negative values less than 1. Raising such values to ever increasing powers forces them to zero. At $\alpha = 0$, however, all functions in (10.5) equal 1. Thus, the limiting function of (10.5) equals 0 everywhere between $-\pi/2$ and $\pi/2$ except at $\alpha = 0$, where its value is 1. The rising exponents in (10.5) bend the curves with such force that, ultimately, they snap.

This example compels a retreat to the "mental bookmark" in Bolzano's proof. Perhaps B is, after all, not continuous. This would cost B all of its charm, which rests on its mix of continuity and non-differentiability. Happily, someone later showed that B *is* continuous; it is not a casualty of over-bending as is the limit of (10.5).

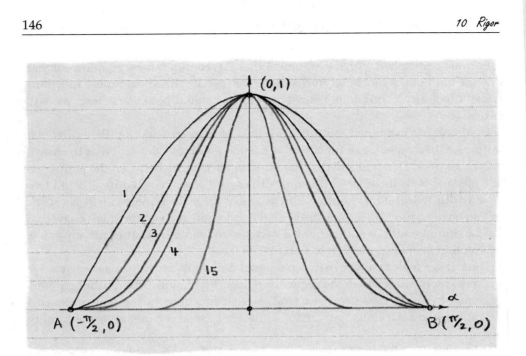

Figure 10.7. The labels on each curve are k in $(\cos\alpha)^k$.

His own intuition prevented Bolzano from believing that he had found a continuous function that had not a single differentiable point. Indeed, he spent years attempting to prove that any continuous function could only fail to be differentiable at a set of points that are isolated from one another. He eventually realized that his function B provided a counterexample to his own search. Later, others showed that B lacks even a single differentiable point.

Karl Weierstrass (Germany, born 1815) refined Bolzano's ideas about the convergence of functions, realizing that there are two kinds; the first, to which Bolzano appealed, is called *pointwise convergence*. Bolzano created B point by point as the limit of sequences of points like $J_1, J_2, J_3, \ldots$ in Figure 10.6. But thanks to example (10.5), we know that functions that pointwise converge do not necessarily pass along the property of continuity to their limit.

Weierstrass distinguished the kind of convergence that *does* transfer continuity as a stronger sort of convergence, and he named it *uniform convergence*. Imagine Bolzano's limiting function B covered by a strip, like a string covered by a strip of cloth; however thin we make this strip, we can find a large enough number m so that all of the stepping stone functions

$$B_m, B_{m+1}, B_{m+2}, \ldots$$

lie entirely in that strip. The discovery of such an m is not simple, primarily because B is so unusual; it can be done, however, and was (many years after Bolzano died).

The algebraic way of describing concepts like continuity and convergence belongs primarily to Weierstrass; the statement (9.12), for example, represents his approach well. His definition of uniform convergence adopts a similar style:

> A sequence of functions $B_1, B_2, B_3, \ldots$ con-
> verges *uniformly* to a function B at a value x
> if for every $\epsilon > 0$, there is a number m so that   (10.6)
>
> $|B_k(x) - B(x)| < \epsilon$ for all $k \geq m$.

The quantity ϵ acts as the width of the strip covering B, and then m marks the boundary beyond which all functions in the sequence are similarly covered. The variable x assumes its values based on the functions in the sequence.

Having distinguished between the two sorts of convergence, Weierstrass invented a function of his own that possesses the peculiar properties of Bolzano's function B. In a way, the function Weierstrass created puts Bolzano's to shame: Bolzano created B by introducing more 'sharp' points at each step, but the functions in the sequence that Weierstrass used have *no* sharp points *ever*. Yet, incredibly, the limiting function of this sequence has *only* non-differentiable points. The closer we look at the boundary between sequences and limits, the more astonishing it seems.

10.4   Dirichlet's nowhere-continuous function

Functions with counterintuitive properties, like those invented by Bolzano and Weierstrass, led some thinkers of the day to lament their discovery. The label 'pathological' — a word that does not mean 'illogical' but, rather, *diseased* — was applied, and it stuck. Unfortunately for the critics, the definitions on which these functions relied for their pathologies were anything but counterintuitive. To this day, we use definitions like (9.12) and (10.6) with full knowledge that they admit to the existence of pathological functions, not because we must use them, but because they are just sort of obvious.

Critics also must contend with the way pathological examples helped to drive the subject toward clarity, not anarchy. Cauchy helped bring clarity to the study of integration by freeing it from its bond to differentiation, a bond made powerful by (7.6), the fundamental theorem of calculus. Later, **Peter Lejeune Dirichlet** (France, born 1805) proposed the function

$$f(x) = \begin{cases} 0 & \text{when } x \text{ is a rational number,} \\ 1 & \text{when } x \text{ is an irrational number} \end{cases} \qquad (10.7)$$

that (productively) put stress on Cauchy's idea of integration. We will see how in a moment.

As he did for differentiation, so too did Cauchy express integration in terms of limits. Consider a function $f(x)$ that is continuous at all values between a and b. Subdivide the horizontal between $a = x_0$ and $b = x_n$ into n parts by choosing

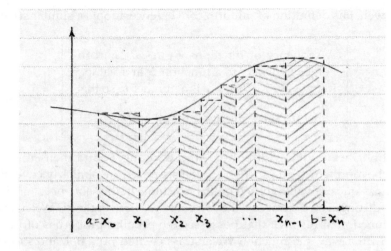

Figure 10.8. Cauchy's non-geometric view of integration can nevertheless be depicted.

values $x_1, x_2, \ldots, x_{n-1}$, and form the sum

$$S = \sum_{k=1}^{n} (x_k - x_{k-1}) f(x_{k-1}) . \tag{10.8}$$

See Figure 10.8, where $n = 8$. Geometrically, we see that S is the sum of the shaded rectangles in the figure. Cauchy, however, made no reference to geometry. Rather than suggest that we make the rectangles more numerous while decreasing their widths, he refers to increasing the number n indefinitely while decreasing the differences $(x_k - x_{k-1})$ in (10.8). He calls the limiting value of S the *definite integral* of f from x_0 to x_n and denotes it

$$\int_a^b f(x) \, dx .$$

We may slightly modify this definition of definite integral to cope with functions that are not continuous, so long as there are a finite number of jump discontinuities.

The definite integral of Dirichlet's function (10.7) is not amenable to Cauchy's method because it fails to be continuous at any of its points. Every rational number r has irrationals indefinitely close to it; that is, for any $\epsilon > 0$, there are irrationals between $r - \epsilon$ and $r + \epsilon$. The same goes for the rationals crowding the irrationals. Thus, if we use (10.1) to test f for continuity at any point, we will discover that the limit does not exist.

Bernhard Riemann (Germany, born 1826) proposed a slightly different definition of integration that could tolerate functions that have an infinite number of discontinuous points in a finite interval. He immediately volunteered a function that was discontinuous infinitely many times within *every* finite interval, then integrated it; this intuition-shattering function encloses an area despite its pervasive discontinuity. Nevertheless, Riemann's approach to integration could not give an answer for Dirichlet's function, which was somehow discontinuous *too often*. This seems to imply that Dirichlet's 'infinite' exceeds Riemann's; but what sense could that make?

10.5 A few final words about the infinite

The story of calculus is rooted in questions about number, geometry, and the infinite. In the 1600s, these threads were woven together, and this conjunction sparked an explosion of results. Criticisms of the underlying logic forced mathematicians to reflect on the foundations of the subject, but attempts to put calculus on a firm footing seemed to provoke controversy as often as they produced solutions. The deeper that mathematicians dug, the more mystery they encountered.

One path we could follow from here starts where section 10.4 ended; the question lurking in the background was this: what is a real number? It may seem absurd to have regressed to this point, but on the contrary, we should celebrate. Any intellectual pursuit worth its salt eventually scrapes its way down to its central vocabulary. In mathematics, we can dig deeper than in most pursuits, and the deeper we go, the more fertile the soil.

Georg Cantor (Germany, born 1845) tilled this soil as fruitfully as anyone. His investigations into the nature of the real numbers not only addressed one of the most commonly overlooked assumptions of the past fifty years (namely, the 'completeness' of the real numbers) but also led to his discoveries about the meaning of 'infinite.' In short, he provided an intuitively pleasing and ironclad proof that there are, in fact, *more* irrational numbers than there are rational numbers. The rationals can be counted — that is, they can be arranged in a list — in a way that the irrationals cannot.

With this idea in the air, **Henri Lebesgue** (France, born 1875) provided yet another technique for integrating that succeeded where Cauchy's and Riemann's had not: his technique *could* integrate Dirichlet's function (10.7). Even better, every function integrable using Riemann's approach is integrable by Lebesgue's.

It seems fitting that the concept of the 'infinite' accompanied us through the story of calculus, as a useful but troubling companion, only to be revealed at the end as infinitely nuanced itself. Scholars do not always respond well to such revelations; Cantor suffered a great deal when his contemporaries unjustly rejected his discoveries. Truth eventually won out over prejudice, and Cantor was vindicated. A few years after Cantor died, **David Hilbert** (Germany, born 1862) declared, "No one shall expel us from the paradise that Cantor has created for us." Hilbert meant not only to thank Cantor for his pioneering work on the infinite, but also to celebrate all of the compelling work that followed from Cantor's results.

One such lovely result is due to **Abraham Robinson**, born in 1918. Robinson lived in so many places that it is difficult to decide upon a country that ought to claim him. He made the 'infinitely small' rigorous, not by appealing to limits, but by inventing a number system that appends *infinitesimals* to the real numbers. These new numbers, which act like the differentials of Leibniz, can be defined precisely and used to recreate all of the results that this book has referenced.

Thus did Robinson loop this story back upon itself, bringing a logical basis to the intuitions shared by thinkers from so many countries and so many centuries. As with every worthwhile intellectual pursuit, calculus is not a tale with a conclusion, but a story that points the way to others.

10.6 Furthermore

10.1 **Riemann rearranges series.** We expended quite a bit of effort in chapter 5 before reaching the sum (5.11) of the alternating harmonic series. It is tempting to believe that a series has been purged of mystery once it has been summed. Riemann reopened the case.

Merely by rearranging its terms, Riemann summed the alternating harmonic series to a value other than $\ln 2$. Take this rearrangement, for example: shift each fraction having an odd denominator to immediately precede the fraction having twice that denominator. The resulting series

$$1 - \frac{1}{2} - \frac{1}{4} + \frac{1}{3} - \frac{1}{6} - \frac{1}{8} + \frac{1}{5} - \frac{1}{10} - \frac{1}{12} + \cdots$$

contains all of the same terms as the alternating harmonic series. However, try grouping some of the neighboring pairs like so:

$$\left(1 - \frac{1}{2}\right) - \frac{1}{4} + \left(\frac{1}{3} - \frac{1}{6}\right) - \frac{1}{8} + \left(\frac{1}{5} - \frac{1}{10}\right) - \frac{1}{12} + \cdots .$$

Simplify, and explain why the sum cannot be $\ln 2$.

Historical note. Riemann discovered a simple way to categorize those series that can be rearranged to give different sums, and those that cannot. His argument would make an excellent point of departure from here.

10.2 **Exploring Bolzano's function.** The fractions $3/8, 5/8$ and so on in Bolzano's function B (described in section 10.2) may not be particularly important. Change them in any way you see fit, and then draw a sketch of B_1, B_2, B_3 like in Figure 10.6.

10.3 **Cauchy and the continuity of the sine function.** Mathematicians excel at translating difficult problems into tractable ones. The transmutation theorem (6.10) epitomizes this strategy; Leibniz calculated the area of a quarter-circle with it as we saw in section 6.4.

When we approximate a function with a series, as in

$$\frac{1}{1+x} = 1 - x + x^2 - x^3 + x^4 - \cdots$$

from section 2.3, we are translating the function, in a way. This translation carries a burden, however: sometimes the series does not converge to the function at every value of x that the function accepts. Some series, like

$$\sin x = x - \frac{x^3}{3!} + \frac{x^5}{5!} - \frac{x^7}{7!} + \cdots , \tag{10.9}$$

do not lose anything in translation; the series converges to the function at all values of x.

On a similar note, it is plausible that when the terms of a series are continuous, and the series converges to a function, then the function will also be continuous. (Echoes of (10.5), however, warn us to step carefully.)

(a) We may use Cauchy's definition of continuity

$$\lim_{h \to 0} [f(x + h) - f(x)] = 0 \qquad (10.10)$$

to check that the terms of the series in (10.9) are continuous. For example, the first term x is a continuous function because

$$\lim_{h \to 0} [(x + h) - x] = \lim_{h \to 0} h = 0 \,.$$

Perform the same check for the next term $x^3/3!$.

(b) What convinces you that every term in the series in (10.9) is continuous?

(c) Cauchy, among others, proved that when continuous functions are added or subtracted, the result is continuous as well. (We will accept this as fact.) Thus, defining s_n to mean that partial sum of the series having n terms, we see that s_n is continuous for all values of n. Cauchy treated the leftover terms as a remainder; if we let s denote the series and r_n denote the terms of s that remain when we pause at s_n, then $s = s_n + r_n$.

If we choose a value of x and find that at some point the remainders r_n stay as small as we wish as n grows without limit, then the partial sums converge at that value of x. The remainders vanish, so to speak.

Cauchy showed that the remainders of the series in (10.9) vanish for *all* values of x. Thus (10.9) is true 'everywhere'; we need not worry about its convergence.

So we have a series composed of continuous terms and continuous partial sums that converge to $\sin x$ at all values of x. With little doubt, the function $\sin x$ is continuous itself.

Cauchy showed this using the identity

$$\sin(A + B) - \sin(A) = 2 \sin(B/2) \cos(A + B/2) \,.$$

Reproduce his argument.

(d) Cauchy then proved that whenever a series of continuous functions converges to a function, then that function is itself continuous. His argument centered on the idea behind his definition (10.10) of continuity: if we compare two arbitrarily close values of x and find that their function values differ by an arbitrarily small amount, then the function is continuous.

A few years after Cauchy published his proof, **Niels Henrik Abel** (Norway, born 1802) published the exception

$$\sin x - \frac{1}{2}\sin(2x) + \frac{1}{3}\sin(3x) - \frac{1}{4}\sin(4x) + \cdots, \qquad (10.11)$$

which meets Cauchy's criteria (each term is a continuous function, and the series converges) but fails to converge to a continuous function. When summed, the terms in (10.11) ultimately 'tear' at $x = \pi$, among other places.

Show that the term

$$\frac{1}{2}\sin(2x)$$

is continuous at all values of x. Can you conclude that all terms in (10.11) are continuous?

Historical note. Cauchy and others pondered exceptions like (10.11) to see which of their assumptions were incorrect. The trouble lay hidden in the difference between a function's continuity near a particular point versus its continuity in a sort of universal sense.

David Bressoud recounts the development of the definition of continuity in *A Radical Approach to Real Analysis*, 78ff. Reading such a book requires some knowledge of calculus at which the last two chapters of this book have only hinted. Pick up any analysis book; almost every one is a sequel to this.

References

The Historical Development of the Calculus (New York: Springer-Verlag, 1979) by C. H. Edwards influenced the content of this book more than any other.

Carl Boyer's article "The History of the Calculus", *The Two-Year College Mathematics Journal* 1 (Spring 1970), 60–86, played a similarly important role in the structure of this book.

Section 1.1

In "Zeno's Arrow, Divisible Infinitesimals, and Chrysippus", *Phronesis* 27 (1982), 239–254, Michael White focuses on the Greek philosopher **Chrysippus** (born c. 280 BCE), who developed Zeno's ideas in a direction that links us to the work of the very last thinker mentioned in this book, Abraham Robinson.

For more on infinite series, see Morris Kline's "Euler and Infinite Series", *Mathematics Magazine* 56 (November 1983), 307–314.

Section 1.2

Greek mathematicians knew that the area of a circle is proportional to its circumference, as explained in C. H. Edwards, *The Historical Development of the Calculus* (New York: Springer-Verlag, 1979), 16–19.

Section 1.3

Some of the geometric details omitted from Archimedes' calculation of the area of a parabolic segment appear in C. H. Edwards, *The Historical Development of the Calculus* (New York: Springer-Verlag, 1979), 35–39. The original argument can be read in T. L. Heath, *The Works of Archimedes* (Cambridge: At the University Press, 1897), 233ff.

The quote by Archimedes on page 7 is from T. L. Heath, *The Works of Archimedes* (Cambridge: At the University Press, 1897), p. 233.

Section 1.4

In contrast with proof by induction, a *combinatorial proof* is a useful deductive method for deriving results in number theory. For a combinatorial proof of a sum that is important in this book, see Arthur Benjamin, Jennifer Quinn, Calyssa Wurtz, "Summing cubes by counting rectangles", *The College Mathematics Journal* 37 (2006), 387–389.

Section 1.5

The history of the notation used by the thinkers in this book appears in Florian Cajoli, *A History of Mathematical Notations* (New York: Cosimo, 2007).

Section 1.6

Philip Straffin, Jr. illustrates ancient Chinese interest in π via the work of a particular mathematician in "Liu Hui and the First Golden Age of Chinese Mathematics", *Mathematics Magazine* 71 (June 1998), particularly on pp. 172–173.

Archimedes approximated π by doubling sides of polygons in a general way that C. H. Edwards demonstrates using the notation of algebra, in *The Historical Development of the Calculus* (New York: Springer-Verlag, 1979), 31–35.

Section 2.1 and 2.2

Victor Katz gives ibn al-Haytham's proofs in "Ideas of Calculus in Islam and India", *Mathematics Magazine* 68, (June 1995), 163–174.

Sections 2.3 and 2.4

Jyesthadeva's argument appears in Ranjan Roy, "The Discovery of the Series Formula for $\pi/4$ by Leibniz, Gregory and Nilakantha", *Mathematics Magazine* 63 (December 1990), 291–306. The original text of Jyesthadeva's argument can be found in *Ganita-Yukti-Bhasa* (Volume 1), translated by K. V. Sarma (Springer, 2008), 183–198.

Further explorations by Islamic mathematicians of the series for $\pi/4$ appear in S. Parameswaran, "Whish's Showroom Revisited", *The Mathematical Gazette*, 76, The Use of the History of Mathematics in the Teaching of Mathematics (March 1992), 28–36.

Robert Young's *Excursions in Calculus* (Washington, D. C.: Mathematical Association of America, 1992), p. 316, sheds light on the connection between Jyesthadeva's formula (2.13) and the odd whole numbers.

Section 2.5

Brahmagupta's *Algebra with Arithmetic and Mensuration*, translated by H. T. Colebrooke (London: John Murray, 1817) contains the problem quoted in exercise 2.1.

Section 3.1

Nicole Oresme's treatise appears, with extensive commentary, in Marshall Clagett's

Nicole Oresme and the Medieval Geometry of Qualities and Motions (Madison, Milwaukee, and London: The University of Wisconsin Press, 1968).

Galileo's observation about uniform acceleration appears in *Dialogue Concerning the Two Chief World Systems* (New York, NY: Dover Publications, 1954), 174.

Sections 3.2 and 3.3

Fermat's treatment of maximums, minimums, and tangents are described in Michael Sean Mahoney, *The Mathematical Career of Pierre de Fermat, 1601–1665* (Princeton: Princeton University Press, 1994). The author reconstructs the flow of Fermat's maturation as a mathematician, and examines the longstanding debate between Fermat and Descartes.

Section 3.3

A survey of the study of tangent lines appears in Ethan Bloch's *The Real Numbers and Real Analysis* (Springer, 2011), 226–230.

Section 3.4

The book by Descartes referenced in this section is translated by David Eugene Smith and Marcia L. Latham in *The Geometry of René Descartes* (New York: Dover Publications, 1954). Pages 95–112 explain how Descartes calculated normal lines, a method which he famously prefaced, "I dare say that this is not only the most useful and most general problem in geometry that I know, but even that I have ever desired to know."

Jeff Suzuki elaborates both on the geometric method of Descartes and on Hudde's helpful algebraic shortcuts in "The Lost Calculus (1637–1670): Tangency and Optimization without Limits", *Mathematics Magazine* 78 (December 2005), 339–353.

Daniel Curtin's "Jan Hudde and the Quotient Rule before Newton and Leibniz", *The College Mathematics Journal* 36 (September 2005), 262–272, explains the observation by Hudde that simplifies the algebra involved in the methods of Fermat and Descartes.

Section 3.5

For more on Mengoli and the convergence of series, see Giovanni Ferraro's *The Rise and Development of the Theory of Series up to the Early 1820s* (New York: Springer, 2009), 8–10.

Oresme's proof that the harmonic series diverges appears in his essay on comparing series; see *A Source Book in Medieval Science* (Cambridge, Massachusetts: Harvard University Press, 1974), 131–135, edited by Edward Grant.

The letter that Descartes wrote (to Claude Hardy) is included in Volume 7 of *Oeuvres de Descartes* (Paris: F. G. Levrault, 1824), 61–65.

Section 4.1

Carl Boyer not only explains Cavalieri's quadrature of the parabola in his article "Cavalieri, Limits and Discarded Infinitesimals", *Scripta Mathematica* 8 (1941), 79–91, but also describes how subsequent mathematicians established his results using number theory.

Section 4.2

Jerry Lodder, in "Curvature in the Calculus Classroom", *The American Mathematical Monthly* 110 (August-September 2003), 593–605, illustrates the importance of the cycloid with two examples: the invention of the tautochrone by Huygens, and the study of elasticity by **Sophie Germain** (France, born 1776).

Section 4.3

Paolo Mancosu goes into great detail about the debate between Guldin and Cavalieri in *Philosophy of mathematics and mathematical practice in the seventeenth century* (New York: Oxford University Press, 1996), 34ff.

Section 4.4

Other results by Kepler are detailed in *The Origins of the Infinitesimal Calculus* by Margaret Baron (Dover Publications, 2004), 108–116. Valerio's argument in exercise 4.3 appears on pages 105–106.

Galileo's paradox can be found in *Dialogue Concerning the Two Chief World Systems* (New York, NY: Dover Publications, 1954), 27–30.

The quote from Zu Geng appears in "The Chinese Concept of Cavalieri's Principle and Its Applications" by Lam Lay-Yong and Shen Kangsheng, in *Historia Mathematica* **12** (1985), 219–228.

Section 5.1

See "Gregory of St Vincent and the rectangular hyperbola" by Bob Burn in *The Mathematical Gazette* 84, (November 2000), 480–485, for a careful explanation of the limiting process that Gregory used in his argument that the quadrature of the hyperbola is connected to arithmetic and geometric series.

Section 5.2

The contributions of de Sarasa toward an understanding of logarithms are described by R. P. Burn in "Alphonse Antonio de Sarasa and Logarithms", *Historia Mathematica* **28** (2001), 1–17.

Section 5.3

Brouncker's argument (and his original diagram) appear in Jacqueline Stedall's *A Discourse Concerning Algebra* (Oxford University Press, 2003), 185–186.

Section 5.4

Ferraro describes Wallis's effort to explain indivisibles using arithmetic, and Mercator's quadrature of the hyperbola on pages 10–20.

In "On the Discovery of the Logarithmic Series and Its Development in England up to Cotes", *National Mathematics Magazine* 14 (1939), 37–45, Josef Ehrenfried Hofmann stresses the role of the logarithmic series in the story of calculus by relating the discoveries of Mercator, Gregory, Newton, and the subsequent refinement of these ideas by **Roger Cotes** (England, born 1682).

Wallis's original arguments appear in Jacqueline Stedall's *The Arithmetic of Infinitesimals: John Wallis 1656* (New York: Springer-Verlag, 2004).

Section 5.5

The development of fractional powers is detailed in Carl Boyer's "Fractional Indices, Exponents, and Powers", *National Mathematics Magazine* 18 (November 1943), 81–86.

Ian Bruce explains how Briggs determined log 2 in "The agony and the ecstasy — the development of logarithms by Henry Briggs", *The Mathematical Gazette* **86** (July 2002), 216–227; the article includes an algorithm for taking square roots and shows where Briggs made a tiny arithmetic error that cascaded through the rest of his calculations.

See page 4 of R. P. Burn, "Alphonse Antonio de Sarasa and Logarithms", *Historia Mathematica* **28** (2001) for other examples like Table 4, and the story of how the logarithm of 1 became associated with 0.

Sections 6.1 and 6.2

For another description of Newton's discovery of the fundamental theorem of calculus, see Hans Niels Jahnke, ed., *A History of Analysis* (Providence, RI: American Mathematical Society, 2003), 76–78. Jahnke also describes Newton's use of *fluxions*, which do not appear in this book.

Section 6.3

The discovery by Leibniz of the fundamental theorem and his transmutation theorem appear in Hans Niels Jahnke, ed., *A History of Analysis* (Providence, RI: American Mathematical Society, 2003), 86–88.

Section 6.4

C. H. Edwards describes the transmutation theorem of Leibniz and its use in calculating the area of the quarter-circle in *The Historical Development of the Calculus* (New York: Springer-Verlag, 1979), 246–248.

Nick Mackinnon provides more historical background to the efforts of Newton and Leibniz to discover the series for $\pi/4$ in "Newton's Teaser", *The Mathematical Gazette* 76 (March 1992), 2–27.

The quote by Isaac Newton is from H.W. Turnbull, et al., eds., *The Correspondence of Isaac Newton* Vol. 2 (Cambridge: Cambridge University Press, 1960), 130.

Section 6.5

Maclaurin's special case of the fundamental theorem of calculus, described in exercise 6.3, is highlighted in Judith Grabiner's "Was Newton's Calculus a Dead End?", *American Mathematical Monthly* 104 (1997), 393–410, which explores whether Maclaurin's role in the development of calculus is underrated. Further, pages 147–148 of Ferraro describe Maclaurin's analysis of inequalities.

Fred Rickey explains Newton's discovery of the generalized binomial theorem (see exercise 6.5) in "Isaac Newton: Man, Myth, and Mathematics", *The College Mathematics Journal* 18 (1987), 362–389.

The sketch of Euler's discovery of a series involving the exponential constant in exercise 6.6 is found in Robert Young, *Excursions in Calculus* (Washington, D. C.: Mathematical Association of America, 1992), 196.

Section 7.1

Hans Niels Jahnke, ed., *A History of Analysis* (Providence, RI: American Mathematical Society, 2003) discusses the new notation of Leibniz on pages 89–90. Figure 7.1 appears on page 89.

Leibniz did not arrive at the product rule in a single try; Frederick Rickey describes his efforts in "The Product Rule", http://www.math.usma.edu/people/rickey/hm/CalcNotes/ProductRule.pdf, 1996.

Section 7.2

On pages 282–288, Baron shows the notation that Leibniz worked with before he developed the symbols that we currently use.

Section 7.3

The quadrature of the cycloid by Leibniz comes from Jahnke, 91.

A translation of Leibniz's derivation of the quadratrix in exercise 7.3 appears in *A Source Book in Mathematics, 1200–1800* (Harvard: Harvard University Press, 1969), edited by D. J. Struik, on pages 282–284.

Section 8.1

David Bressoud explains the origins of the term *sine* in "Was Calculus Invented in India?", *The College Mathematics Journal* 33 (January 2002), 2–13.

Section 8.2

The derivation of the series for sine and cosine is due to David Bressoud, *A Radical Approach to Real Analysis* (Washington, D.C.: Mathematical Association of America, 2007), 9–12, who also explains how the ancient Greeks managed to compile an accurate table for the trigonometric functions.

Section 8.3

Newton's derivation of the series for $\sin \alpha$ from his series for $\arcsin \alpha$ is detailed in William Dunham's *The Calculus Gallery* (Princeton, NJ: Princeton University Press, 2005), 16–18.

Section 8.4

The outline of Euler's sum of the reciprocals of the squares is thanks to Judith Grabiner in "Who Gave You the Epsilon? Cauchy and the Origins of Rigorous Calculus", *American Mathematical Monthly* 90 (1983), 187.

Section 9.1

The description of D'Alembert's investigations of convergence appears in David Bressoud, *A Radical Approach to Real Analysis* (Washington, D.C.: Mathematical Association of America, 2007), 41–43.

Section 9.2

The meaning of derivative evolved in much the same way as did the function, and Judith Grabiner provides an overview in "The Changing Concept of Change: The Derivative from Fermat to Weierstrass", *Mathematics Magazine* 56 (September 1983), 195–206.

Section 9.3

Lenore Feigenbaum carefully explores the origin of Taylor's method in her dissertation *Brook Taylor's 'Methodes Incrementorum'* (Yale University, 1981).

Section 9.4

Florian Cajori provides an overview of efforts from Leibniz to Cauchy to come to grips with limits, and how well limits clarify the mysteries of the infinitely small and infinitely many, in "Grafting of the Theory of Limits on the Calculus of Leibniz", *The American Mathematical Monthly* 30 (July-August 1923), 223–234.

Cauchy's version of (9.13), the definition of convergence of a series, appears in *Cauchy's Cours d'analyse*, translated by Robert E. Bradley and C. Edward Sandifer (Springer, 2000), 86–87.

Section 10.1

Israel Kleiner describes the tension between geometric and algebraic models in "Evolution of the Function Concept: A Brief Survey", *The College Mathematics Journal* 20 (September 1989), 282–300.

Cauchy's definition of continuity appears in *Cauchy's Cours d'analyse*, translated by Robert E. Bradley and C. Edward Sandifer (Springer, 2000), 26. Cauchy relaxes the stipulation that the "neighborhood" on which f is continuous must have radius α.

Section 10.2

A translation of Bolzano's original description of his function B appears in Steve Russ, *The Mathematical Works of Bernard Bolzano* (Oxford: Oxford University Press, 2004).

Vojtěch Jarník celebrates Bolzano's discovery of his function B in *Bolzano and the Foundations of Mathematical Analysis* (Praha: Society of Czechoslovak Mathematicians and Physicists, 1981), 33–42, but believes that the main significance of Bolzano's work lies in his careful treatment of continuity and differentiability.

The quote on page 140 by Cauchy is from *Résumé des leçons données à l'Ecole Royale Polytechnique sur le calcul infinitésimal, Oeuvres* Ser. 2, Vol. 4 (Paris: Gauthier-Villars, 1899), and appears in Edwards, 313.

Section 10.3

Proofs that Bolzano's function B is non-differentiable everywhere and that B is the uniformly continuous limit of Bolzano's sequence of functions appear in Vojtěch Jarník's *O funci Bolzanově* [*On Bolzano's function*] (Časopis Pěst. Mat. 51 (1922), 248–266.

The brief treatment of Weierstrass is greatly expanded in William Dunham's *The Calculus Gallery* (Princeton, NJ: Princeton University Press, 2005), 128ff.

Section 10.4

William Dunham explains Riemann's example of a Riemann-integrable function that is discontinuous infinitely many times within every finite interval in *The Calculus Gallery* (Princeton, NJ: Princeton University Press, 2005), 107–112.

Citations for chapter-opening quotes

2. Thom Satterlee, "Ibn Khatir Tells How He Survived the Black Death", *Burning Wyclif* (Texas Tech University Press, 2006).

3. Ellen McLaughlin, "Circe and the Hanged Man", *Penelope*, New Amsterdam Records, 2010.

5. David Byrne, "Finite = Alright", *Feelings*, Luaka Bop, 1997.

Index

About the Author

David Perkins earned his doctorate in mathematics at the University of Montana (advised by P. Mark Kayll). He taught at Houghton College for ten years before moving to Luzerne County Community College to live and work with his wife Michelle LaBarre. In 2004, he became an 'orange dot' as a member of Project NExT.

PARASITES

Leeches

Other titles in the Parasites series include:

Chiggers
The Ebola Virus
Fleas
Guinea Worms
Hookworms
Lice
Salmonella
Tapeworms
Ticks

Leeches

Sheila Wyborny

KIDHAVEN PRESS

An imprint of Thomson Gale, a part of The Thomson Corporation

THOMSON

GALE

Detroit • New York • San Francisco • San Diego • New Haven, Conn. • Waterville, Maine • London • Munich

LIBRARY OF CONGRESS CATALOGING-IN-PUBLICATION DATA

Wyborny, Sheila, 1950–
 Leeches / by Sheila Wyborny.
  p. cm. — (Parasites)
 Includes bibliographical references (p.   ).
 ISBN 0-7377-3050-1 (hardcover : alk. paper)
 1. Leeches—Juvenile literature. I. Title. II. Series.
 QL391.A6W93 2005
 592'.66—dc22

 2004017025

Printed in the United States of America

CONTENTS

CHAPTER 1

Getting to Know Leeches

Leeches are **parasites** that survive by feeding off the blood and other body fluids of **hosts**. Any mammal, reptile, snail, worm, fish, and even some insects can be a host for a leech. Hundreds of species of leeches, most ranging in length from ½ inch to 6 inches (1cm to 15cm), have been identified. Some leeches are brown, while others are green, orange, red, or even dark yellow. But despite their differences in size and color, these species have many similarities.

Leeches are paras[itic] feed on blood. S[hown] here is a magnif[ication] of a leech's bloo[d] mouthparts.

Animal Family

All leeches belong to an animal group, or phylum, called **annelids**. The other main members of the annelid phylum are earthworms and sea worms. Although all annelids do not look alike, they have some of the same body features. For example, annelids are **invertebrates**. This means that they do not have a hard skeleton.

Looking at Leeches

All leeches, both large and small, have thirty-four body segments. They are also the only annelids with suckers. Leeches have a sucker on each end of their bodies. The head-end sucker is for feeding, and the tail-end sucker is for hanging onto a host. Leeches also have very sensitive organs called **segmental receptors**. These organs can sense the presence of other animals or even chemicals in their environment, as well as the slightest movement in water and the tiniest vibrations on land. Leeches also have organs similar to eyes. Between two and ten of these light-sensitive organs are located toward the head of the body.

An Efficient Parasite

1. Head-end sucker is for feeding. Some leeches, like the American medicinal leech, have rows of sharp teeth (inset).

2. All leeches have body segments. These segments look like rings around the worm's body, and help the animal to move and hold its shape.

3. Tail-end sucker helps the leech cling to its host while feeding.

Like the earthworm and sea worm, leeches have soft segmented bodies. These segments look like rings around the body. The segments contain liquid that helps annelids move and hold their shape.

Three Leech Groups

Leeches are categorized into three groups according to the way they feed. Each group has a different type of mouth and so feeds differently. The first group, including the American medicinal leech, has a round mouth with jaws of hundreds of tiny conical teeth. A medicinal leech can have up to three jaws of teeth. The second group, which includes the Amazon leech, inserts a needlelike tube, called a **proboscis**, into the flesh of the host. Both types of leeches secrete liquids into the wound, which numb the host's skin and prevent the blood from clotting. This makes it easier for the leech to suck the blood. The third group of leeches, called worm leeches, have no jaws, teeth, or proboscis. They swallow small prey whole, such as tiny worms and snails.

Life Cycle of a Leech

No one is sure exactly how long leeches live, but most scientists believe that some leeches can live for several years. During their lifetime, they reproduce two or three times. Leeches have both male and female reproductive organs and carry both eggs and sperm. Leeches do not fertilize their own eggs, however. Two adult leeches mate by exchanging sperm. The eggs and sperm come together inside each parent leech's body.

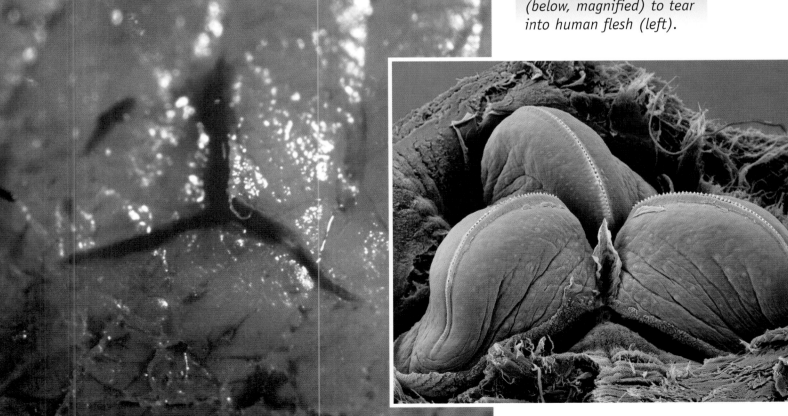

An American medicinal leech used its sharp teeth (below, magnified) to tear into human flesh (left).

In the spring, the fertilized eggs are laid in a nutrition-rich fluid inside a soft saddlelike **cocoon** that surrounds some of the segments on the parent's body. The parent struggles out of the cocoon and seals both ends with a mucus plug. After escaping the

Leeches deposit their eggs in a protective cocoon (left). Below, baby medicinal leeches ooze out of their cocoon.

cocoon, the leech presses the soft cocoon against a hard surface, such as a log or a rock, where the outside of the cocoon hardens into a solid surface. In most types of leeches, this is the end of the parents' work.

A dozen or more eggs feed on the yolklike fluid and grow inside the cocoon. Depending on the type of leech, it can take from a few weeks to several months for the eggs to hatch and for the baby leeches to leave the cocoon. When they hatch sometime in the summer, baby leeches, a ½ inch (1cm) or less in length, look just like their parents. They have all body parts but are much tinier. With no adult leeches around, these baby leeches are completely on their own. By the following spring, they are adults and ready to mate and lay eggs. Some leeches are eaten by predators, such as fish and turtles. Many baby leeches, however, survive to become predators themselves.

Living with Leeches

Opposite: *A terrestrial leech waits for a host on a plant leaf. Leeches let their meals come to them.*

L eeches are found in many countries throughout the world. In some places they can be a real danger to both people and animals. They do not search for food or chase their prey. They simply wait patiently and let their meals come to them.

Lurking Leeches

The leeches' sense organs, segmented receptors, play a very important role when leeches are ready to feed.

A hungry leech is on full alert as it waits, its body fully extended and completely motionless. By extending its body, the leech makes its segmental receptors extremely sensitive. They can sense light, motion, body oils, blood, and even carbon dioxide exhaled by humans and animals anywhere in the area.

When the leech senses a host approaching, it wriggles until it comes into contact with the host. It then attaches itself to the host with its suckers. Sometimes, the host does not even know that a bloodsucking hitchhiker is on board.

Land and Water

Most leeches live in freshwater lakes, marshes, shallow ponds, rivers, and streams. Not liking strong currents or light, they can usually be found in shady areas or under rocks, logs, or other debris on the bottom of slow-moving or stagnant bodies of water. During the day, they remain still and hidden from the light. At night, those not resting from feeding are more active, inching along the bottom or swimming.

Most leeches, such as these medicinal leeches, live in freshwater.

Some leeches, however, are **terrestrial**. This means they live on land. Some terrestrial leeches live on low-growing plants or in the moist soil found in places such as rain forests near the equator, where the humidity is high. They stay out of the light by attaching to the bottoms of leaves or under debris on the forest floor. If the surface soil begins to dry, some leeches can survive by burrowing under the soil and remaining there until it rains. Becoming dry and rigid, they appear to be dead. Within ten minutes of rainfall, however, the leeches absorb water and become active again. In addition to freshwater and terrestrial leeches, a few types of marine leeches live in the Atlantic Ocean and feed on fish.

From Small Problems to Serious Danger

Wherever they live, leeches can cause problems. Throughout history, leeches all over the world have been the subject of horror stories, some true and some false. In fact, some people call leeches the vampires of the worm world.

In the United States, problems with leeches are usually minor. Leeches sometimes attach themselves to people swimming in freshwater lakes or to hikers cutting across small streams. Sometimes, the people who pick up these bloodsucking hitchhikers are not aware of them until the leeches are engorged with blood, swollen to the size of small bananas. This is because of the enzyme in the leech's mouth that deadens the person's skin. In most cases, though, leeches do not take enough blood to cause a person harm.

This is not the case, however, in some tropical parts of the world. In warmer regions, leeches can cause seri-

A man's leg bleeds (right) from leech bites (below). Leeches rarely draw enough blood to cause harm.

ous health problems, even death. In 1799, for instance, soldiers of French emperor Napoléon were serving in Egypt. There was little water in the desert for the soldiers to drink, so they drank from whatever water sources that were available. Leeches lived in some of this water. When the soldiers drank, they swallowed baby leeches. The leeches attached themselves to the soldiers' mouths and throats and began feeding. As they fed, the leeches became larger. Many soldiers died of suffocation when the swollen leeches closed off their throats. Other soldiers died from blood loss, because so many leeches attacked them. In more recent times, soldiers in Vietnam, in Southeast Asia, became seriously ill from leeches picked up on jungle hikes.

Unlike these medicinal leeches, some tropical leeches can cause serious health problems.

Some animals also suffer from leech attacks. A leech in the genus Theromyzon feeds on blood from the nasal passages, throat, and eyes of waterfowl. Ducks, for instance, are sometimes blinded and have

Although leeches can be very dangerous, the threat of these parasites can be controlled.

trouble breathing because of the leeches. Affected ducks can be identified because they breathe with their bills open. The ducks can be successfully treated, however, by having the leeches removed from their nostrils with forceps. But despite the problems leeches cause for humans and animals, the damage they do can be controlled.

CHAPTER 3

Making War and Peace with Leeches

As with other parasites, the world will never be completely rid of leeches. However, even if it could be done, eliminating them would be a bad idea. Although they can be pests, leeches are a part of the food chain. Leeches are eaten by larger animals, and in turn, some of these larger animals are eaten by people. Breaking any part of the food chain can harm other animals. It is best to avoid leeches and the places where they live, but it is also possible to control them.

Avoiding and Controlling Leeches

Since leeches like standing or stagnant water, swimmers and hikers should avoid these places whenever possible. If wetlands cannot be avoided, hikers can wear high-top walking boots when traveling along trails in areas where shallow water covers the trail. It is easier to remove a leech from a boot than from the skin. Also, swimmers who choose to swim in lakes and ponds should enter the water from platforms or boats rather than from shallow areas, where leeches wait on the bottom.

Communities with large areas of stagnant water or swamps cannot completely avoid leeches. In these places, other measures have to be taken to prevent problems. One way is called bait trapping. One-pound coffee cans are drilled with small holes, baited with meat, and dropped with lines into the water. The leeches crawl inside the can and attach themselves to the meat. When the cans are lifted, they are filled with leeches. These cans are then disposed of in a special place. Another way to control leeches, especially in farm ponds and small lakes, is with a chemical substance called **copper sulfate pentahydrate**. This

Opposite: A scientist examines a jar of captive-bred leeches. Leeches like these can be used as part of many medical treatments.

treatment will kill the leeches in a pond or lake. The chemical does not kill the leech eggs, however. Several treatments are needed to completely eliminate leeches from the lake or pond.

Putting Leeches to Work

Some leeches, such as the American medicinal leech, have been put to good use. Leeches are sometimes used to help reattach small body parts. In 1977, a nine-year-old boy's ear was bitten off by a large dog. After the ear was surgically reattached, leeches were placed on the wound to help increase blood flow to the ear. The treatment was successful. Similar treatments have been used to help restore blood circulation to reattached fingers and toes. The suc-

A medicinal leech is used to restore blood circulation to a toe surgically reattached to a man's foot.

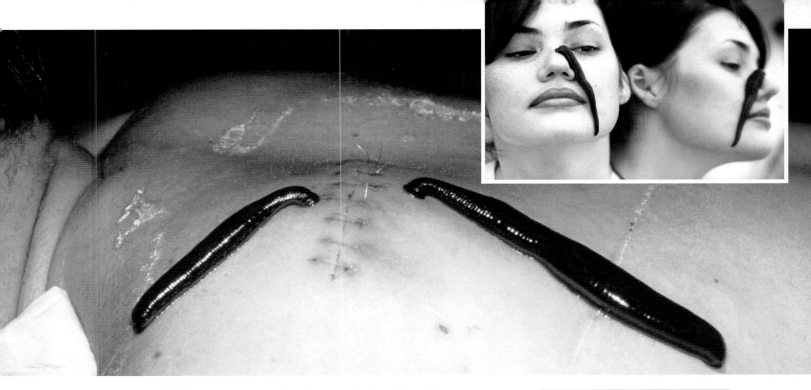

cess of this treatment is due to hirudin, the **enzyme** leeches secrete into wounds when they bite. It is an **anticoagulant**, which increases blood flow by preventing blood from clotting.

In some cases, leeches have been found to reduce arthritis pain. Arthritis causes joints to hurt and swell. Hirudin has been successful in reducing inflammation, particularly in knees. To do this, doctors use the

Doctors sometimes use leeches to heal blood clots such as those on this man's back and this woman's face.

same method that healers have used for thousands of years. The leeches are attached to the knee area to feed. As they feed, the enzyme enters the tissues and acts directly on the inflamed areas, reducing the tenderness and swelling.

While some scientists and doctors continue to research new ways to use leeches to ease suffering, other scientists continue to search for ways to eliminate leeches in parts of the world where they are a health problem. As with any living creature, scientists will have to be very careful with methods they develop to control large populations of leeches since both people and the delicate balance of nature must be protected.

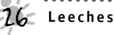

annelids: Wormlike animals with long, cylindrical, segmented bodies.

anticoagulant: A substance that prevents blood from clotting.

cocoon: A protective fibrous case that covers the larvae of certain kinds of insects.

copper sulfate pentahydrate: A chemical used to kill leeches in lakes and ponds.

enzyme: A protein produced by living animals that causes certain chemical reactions.

hosts: Animals or plants on which other animals live and from which they get nourishment.

invertebrates: Animals with no backbone.

parasites: Organisms that feed off other living organisms.

proboscis: A hollow feeding tube found in some invertebrates and used for feeding and sucking.

segmental receptors: A leech's sense organs.

terrestrial: A way to describe anything that lives on land.

FOR FURTHER EXPLORATION

Books

Beth Blaxland, *Annelids: Earthworms, Leeches, and Sea Worms.* Philadelphia, PA: Chelsea, 2002. This book describes the annelid phylum, including leeches. It is illustrated with a number of close-up photographs.

Cheryl Mays Halton, *Those Amazing Leeches.* Minneapolis, MN: Dillon, 1989. This book details the life cycle of leeches, ways this parasite causes problems, and medicinal uses.

Web Sites

Access Excellence: The National Health Museum (www.accessexcellence.org/LC/SS/leechlove.html). This Web site includes a section on the care and feeding of leeches. The reading level is appropriate for elementary school students.

Australian Museum Online (www. austmus.gov.au/factsheets/leeches. htm). This Web site provides fact sheets on many living organisms found in Australia. The fact sheet on leeches deals with the physiology, types, locomotion, habitats, and uses of leeches in medicine. Both the writing and the diagrams are at an appropriate level for elementary school students.

King County Washington Web Site (http://dnr.metrokc.gov/wlr/waterres/Bugs/Leeches.htm). This Web site, written in a student-friendly style, provides descriptions of leeches, their behaviors, feeding, and importance to medicine.

INDEX

Sheila Wyborny, a retired teacher, has written nonfiction books for children for the past four years. She lives in a small airport community near Houston, Texas, with her husband, an engineer. Having their Cessna aircraft literally in their backyard has enabled the Wybornys to have many fun, spur-of-the-moment flying adventures.